Synthesizers and Subtractive Synthesis 2

Synthesizers and Subtractive Synthesis 2

Application and Practice

Jean-Michel Réveillac

iSTE

WILEY

First published 2024 in Great Britain and the United States by ISTE Ltd and John Wiley & Sons, Inc.

ISTE Ltd
27-37 St George's Road
London SW19 4EU
UK

www.iste.co.uk

John Wiley & Sons, Inc.
111 River Street
Hoboken, NJ 07030
USA

www.wiley.com

Library of Congress Control Number 2023951659

British Library Cataloguing-in-Publication Data
A CIP record for this book is available from the British Library
ISBN 978-1-78630-925-9

Contents

Preface

If you want to know if this book is for you, how it is constructed and organized, what is in it, and what conventions will be used, you have come to the right place, this is the place to start.

Target audience and prerequisites

This book is intended for all those who are interested in sound synthesis and synthesizers, whether they are amateurs or professionals, or even musicians, performers or composers.

The information presented in some sections requires basic knowledge of general computing and digital audio.

For some work on microcomputers, you will need to have good knowledge of the operating system (paths, folders and directories, files, names, extensions, copies, moves, etc.).

For exercises based on the VCV Rack and Native Instruments Reaktor Blocks software synthesizers, you will need to know their philosophy, general principles of design and use in order to build a VCV or Reaktor Software Modular Rack.

For exercises related to the visual programming languages Max/MSP and Pure Data, basic knowledge of their interface and the commands of their editors will be necessary.

If you do not feel comfortable with these prerequisites, a set of books and tutorials are mentioned in the reference section of this book, which will help develop your knowledge.

Possession of a synthesizer based on subtractive synthesis will be a plus, especially if it is an ARP2600, a Minimoog, a Novation Bass Station II, a Behringer Neutron or an Arturia MatrixBrute. Software or hardware clones of these machines are also welcome.

Software such as Pure Data or VCV Rack can be downloaded easily and for free, as can software clones of some synthesizers (Minimoog, ARP 2600). Consult the links in the reference section of this book for this purpose.

Organization and contents of the book

This book consists of two volumes:

1) *Synthesizers and Subtractive Synthesis 1: Theory and Overview.*

2) *Synthesizers and Subtractive Synthesis 2: Application and Practice.*

Volume 1 successively presents a preface, specifying the contents and the writing conventions used, and then an introduction followed by five chapters, a conclusion and two appendices:

– sound synthesis;

– different types of synthesis;

– components, processing and tools;

– work environment;

– CV/Gate and MIDI.

The conclusion, as its name suggests, attempts to assess the current state of subtractive synthesis and synthesizers.

Appendices 1 and 2 provide some additional elements and some reminders. You will find information in the following order:

– General MIDI 1 and 2 instruments;

– MIDI boxes, mergers and patchers.

Volume 2 presents a preface identical to that of Volume 1, followed by five chapters, a conclusion and four appendices:

– subtractive synthesis, the beginnings;

– subtractive synthesis, the fundamentals;

– advanced subtractive synthesis;

– duophony, paraphony and polyphony;

– sequencer and arpeggiator.

Appendices 1– 4 provide some additional information in the following order:

– USB connectivity;

– Pure Data extensions;

– keyboards and interface;

– MIDI notes, numbers and frequencies.

The conclusion sheds light on the contents of the book and a brief overview of the future evolution of sound synthesis systems and software.

At the end of this book, you will find references and a list of internet links.

A glossary is also present, and it will explain certain acronyms and terminology very specific to sound synthesis and synthesizers.

Each of the chapters can be read separately. If concepts that are dependent on another chapter are present, the references to the relevant sections are indicated. However, Chapter 1, devoted to sound synthesis, provides the necessary foundations for understanding the subsequent chapters.

If you are a new reader on the subject, I strongly advise you to read Chapter 1 first; the following chapters will then be clearer.

For everyone else, I hope you will discover new notions that will enrich your knowledge.

Conventions

This book uses the following typographical conventions:

– *Italics*: reserved for important terms used for the first time in the text, which may be present in the glossary at the end of the book, mathematical terms, comments, equations, expressions or variables.

– UPPER CASE: reserved for command names, entry, exit, or connection points, specific functions, modules belonging to the different hardware or software synthesizers used in the exercises. It can also be elements, options or choices within menus present in the interface of a program.

– `Courier` font: reserved for objects manipulated within the visual programming software Max/MSP and Pure Data.

Notes are indicated by the presence of the keyword:

NOTE.– These notes complete the explanations already provided.

Figures and tables all have a description which is often useful for understanding.

Vocabulary and definitions

As with all techniques, subtractive sound synthesis and synthesizers have their own vocabulary, with words, acronyms, abbreviations, initials and proper nouns not always familiar. This is the role of the glossary already mentioned above.

Acknowledgments

I would particularly like to thank the ISTE team, and my editor Chantal Ménascé, who trusted me.

Finally, I would like to thank my wife, Vanna, and my friends, passionate about the subject, who supported me throughout the writing of this book.

January 2024

Introduction

Where volume one gathered information and theoretical knowledge, this volume brings together practical exercises carried out on hardware or software synthesizers of several categories: wired, semi-modular, or modular.

The equipment chosen is the Behringer 2600 or the ARP 2600, the Minimoog, the Novation Bass Station II, the Behringer Neutron and the Arturia MatrixBrute, for hardware synthesizers. On the software side, the modular VCV Rack and Native Instruments Reaktor complete the list. A large part of the exercises is reserved for the Max/MSP and Pure Data visual programming environments.

Figure I.1. *The five machines covered in this book. For a color version of this figure, see www.iste.co.uk/reveillac/synthesizers2.zip*

I wanted to put my exercises within the reach of as many people as possible by choosing some affordable machines (Neutron, Bass Station II) in terms of cost and two free open-source software (VCV Rack, Pure Data).

As far as software is concerned, this book is not a learning tutorial. I believe the reader will already know the fundamental bases to carry out each of the examples.

As for the equipment, the user manuals for each machine will provide you with the necessary elements.

If you have no experience, refer to the reference section at the end of the book, where you will find various links to access all the documentation and downloadable tutorials.

The exercises are ordered with increasing difficulty, but there is nothing to prevent you from completing them in the order you want. However, the elements already covered are not recalled for each exercise; they are assumed, so you may need to go back to Volume 1 to review some of them.

Depending on their specificities, some exercises do not need to be presented for all synthesizers, hardware, software or languages covered in this book.

Chapters 1 and 2 focus on the key elements of a monodic subtractive synthesizer, oscillator, filter, envelope generator, low-frequency generator and noise generator.

Chapter 3 covers more advanced features, available only on some synthesizers: ring modulation, sample and hold, and sound effects.

Chapter 4 responds to a subject that has long inconvenienced synthesizer manufacturers because its implementation, until the mid-1980s, required sophisticated and expensive electronics, that of polyphony.

Chapter 5 is not really related to subtractive synthesis, but rather to shaping and manipulation tools: sequencers and arpeggiators.

For all exercises, when equipment is used, this is indicated (in cyan for controls and red for everything else) on each of the figures depicting the synthesizer front panel.

To visualize what each parameter does, the user-editable controls are shown in cyan, and controls that the user must position exactly as indicated are in red. The uncolored commands have no influence on the sound reproduction for the given exercise.

For exercises using hardware or software without an integrated keyboard, a MIDI keyboard and an audio interface or a virtual keyboard can be used. Be sure to set them correctly (MIDI channel, port, type of interface, etc.). Appendices 2 and 3 will provide some explanations to novices in this field.

The exercises will sometimes be followed by more advanced work, interspersed with examples using Max/MSP, Pure Data and VCV Rack software. They complement and enrich the exercises by giving you new avenues of research and development.

With these remarks, it is time to get hands-on with the machines, put together modular synthesizers, and open our minds and especially our ears!

1

Subtractive Synthesis: The Beginning

To simplify the search, I have summarized in Table 1.1, for each of the exercises in this chapter, the hardware and software chosen and the sections where you can find them.

Exercise	Hardware/software	Section
No. 1. 1 VCO, 1 VCA	Behringer Neutron	1.1.1
	Behringer 2600 (ARP2600)	1.1.2
	Max/MSP	1.1.3
	Pure Data	1.1.4
	VCV Rack	1.1.4
No. 2. 1 VCO, 1 VCA, 1 EG	Behringer Neutron	1.2.1
	Behringer 2600 (ARP2600)	1.2.2
	Max/MSP	1.2.3
	Pure Data	1.2.4
	VCV Rack	1.2.5
No. 3. 1 VCO, 1 VCA, 1 EG, 1 VCF	Behringer Neutron	1.3.1
	Max/MSP	1.3.2
	Pure Data	1.3.3
	VCV Rack	1.3.4

Table 1.1. *Summary of the different exercises*

Before starting, it seems appropriate to clarify that the objective of the exercises, in all chapters, is to learn about subtractive sound synthesis in order to understand it better and master it, but also to allow your ears to discover the sensations and the sounds reproduced by each of the signals that you are going to create. What is the point of doing sound synthesis if not to discover new sounds or understand what it means to reproduce a sound, a timbre, or an already existing tone such as that of a musical instrument or a familiar noise? We need to remember to listen rather than hear, pay attention and discover the richness of the sound universe surrounding us by synthesizing our main constituents.

NOTE.– As the exercises are presented, specific comments will not be repeated to avoid overloading the explanations. In case of doubt or misunderstanding, I advise the reader to go back and read the previous exercises to find the explanatory elements that may be missing.

1.1. Exercise 1 – generate sound with a single oscillator

The basics of sound synthesis are to generate sound from an oscillator. A classic configuration contains an oscillator (VCO) and an amplifier (VCA), which can be preceded by an input controller, often a keyboard.

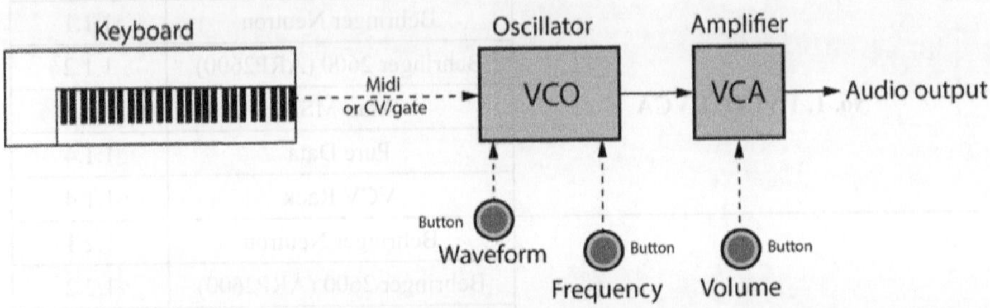

Figure 1.1. *Subtractive sound synthesis with an oscillator. The command signals are represented with dotted lines*

1.1.1. *Behringer Neutron*

The Behringer Neutron semi-modular synthesizer (see Volume 1, section 4.1.3) has two oscillators whose signals can be mixed. For this exercise, we will use Oscillator 1 (Oscillator 2 would have worked the same way, as these two oscillators are identical) and a patch cable, on the matrix, to route it to the VCA. The oscillator has five waveforms, available over three octaves.

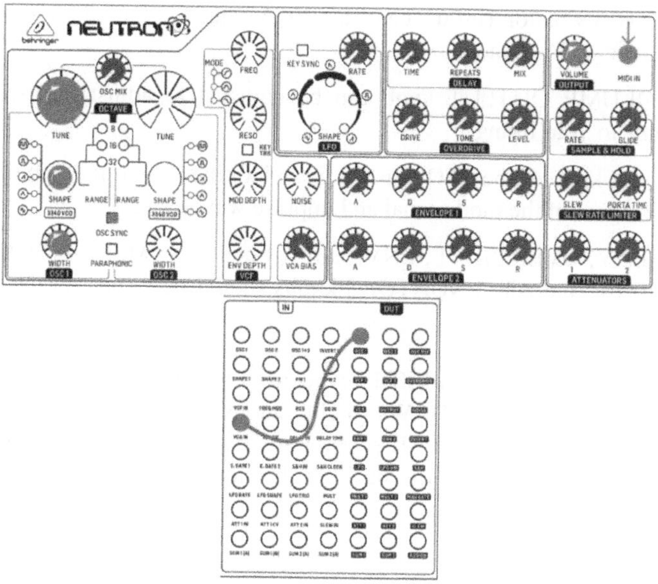

Figure 1.2. *Exercise 1 for the Behringer Neutron. For a color version of this figure, see www.iste.co.uk/reveillac/synthesizers2.zip*

Instructions and comments are as follows:

– connect a MIDI keyboard to the MIDI IN input (on the front panel) and switch it on. Check that the keyboard and the Neutron are placed on the same MIDI channel number (1–16). On the Neutron, the adjustment is made with the DIP switches located on the back;

– connect the audio output (OUTPUT) on the back to an amplification system;

– switch on the Neutron;

– connect the output (OUT) of oscillator 1 (OSC1) with a patch cable to the input (IN) of the VCA (VCA IN) on the input–output matrix to the right; this will route the signal produced by the oscillator directly to the amplifier input;

– the controls of oscillator 1 (OSC1) are all accessible. We find the agreement (TUNE), the waveform (SHAPE: mode-tonal, square-pulse, sawtooth, triangle, sinusoid) and the width (WIDTH), which modifies the width of the pulse or the tonal mode when chosen as the waveform and the octave (RANGE – 8', 16' or 32');

– the oscillator mix control knob (OSC MIX) must be turned all the way to the left;

– the VC BIAS button (influence of the VCA) must be turned all the way to the right to open the VCA to the maximum;

– the VOLUME OUTPUT knob can be adjusted to adjust the overall audio output volume to your liking;

– the GATE knob of the LFO must be turned all the way to the left; otherwise, it modifies the signal of the mode-tonal and impulse waveforms;

– all DELAY, OVERDRIVE, ENVELOPE 1 and 2, SAMPLE & HOLD, SLEW RATE LIMITER and ATTENUATORS knobs should be turned all the way to the left.

NOTE.– Instead of connecting a MIDI keyboard, it is possible to use a virtual MIDI keyboard or the one in some DAWs (digital audio workstations). In this case, the connection is made via the USB-MIDI ports of the microcomputer and the Neutron. For the latter, it is located at the back of the device. Again, the MIDI channel numbers must be identical.

1.1.2. *Behringer 2600 (ARP 2600)*

The Behringer 2600 is one of the ARP 2600 clones (see Volume 1, section 4.1.1) and has similar, if not superior, functionality to today's synthesizers. It is semi-modular but has no separate matrix, and the connection points are distributed on each front panel module.

For this exercise, we will need a patch cable connected to the VCA, the other end of which can be moved to obtain the various signals generated by the oscillator since the equipment does not have a waveform selector.

Oscillator 2 was chosen because it has four waveforms; oscillator 3 would have provided the same results, with these two oscillators being equivalent.

The audio output of the 2600 is stereophonic by default, so you can choose to use both channels or just one. The slide control, PAN, manages the left–right balance.

NOTE.– The names of the commands are not always specified on the serigraphy of the front panel of the ARP 600, Behringer 2600, Korg 2600 or other clones. You can refer to these in the figures for each exercise.

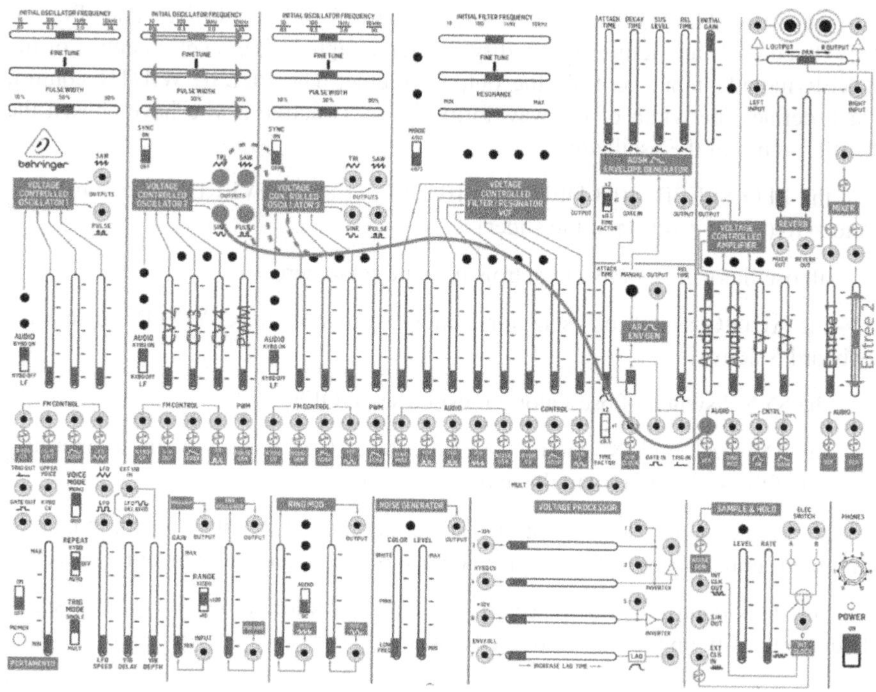

Figure 1.3. *Exercise 1 for Behringer 2600 (ARP 2600). For a color version of this figure, see www.iste.co.uk/reveillac/synthesizers2.zip*

Instructions and comments are as follows:

– connect a MIDI keyboard to the MIDI IN input on the back and switch it on. Check that the keyboard and the 2600 are set to the same MIDI channel number (1–16). On the Behringer 2600, the adjustment is made using the DIP switches on the back. If you are on an ARP 2600, a KORG 2600, or another clone, use the appropriate keyboard;

– connect the one or two audio outputs (L OUTPUT – R OUTPUT) located at the front to an amplification system;

– switch on the 2600;

– connect, using a patch cable, one of the four outputs, TRI, SAW, SINE, or PULSE to input 1 of the VCA;

– the INITIAL OSCILLATOR FREQUENCY, FINE TUNE and PULSE WIDTH controls are active and modify the audio signal. The level 2 control of the MIXER module input controls the output volume. The PULSE WIDTH control, which defines the width of the waveform square, is active if you have chosen the PULSE output of the oscillator;

– the LFO SPEED, VIB DELAY, VIB DEPTH and REVERB controls must be set to zero (lower position of the slider);

– all oscillator 2 input level controls must be set to zero;

– the AUDIO switch of oscillator 2 must be placed on KEYB ON so that the keyboard is active;

– the SYNC switch of oscillator 2 must be set to OFF;

– the INITIAL GAIN control must be placed at its maximum (uppermost position of the slider) to obtain the optimal gain of the VCA;

– the audio signal level control 1 entering the VCA must be placed at its maximum (uppermost position of the slider) to obtain a maximum incoming level;

– the PAN control can be placed in the center position to obtain an audio signal of equal intensity on the right and left outputs.

1.1.3. *Max/MSP*

The Max/MSP exercises in this book have been tested and designed with version 8.5.4, they may also work on different versions but noticeable differences may occur in some cases[1].

You will find in this exercise some of the main Max/MSP functions.

NOTE.– In all Max/MSP exercises, we will not use Live objects[2] (Ableton) to allow all users to carry out the work.

1 Available at: www.cycling74.com/.

2 Live is a software music sequencer developed by Ableton that has been available since 1999. Since 2009, Ableton and Cycling ’74, the distributor of Max/MSP, have been working together and have developed “Max for Live”, a version of Max, integrated with Live. In 2017, Ableton acquired Cycling ’74.

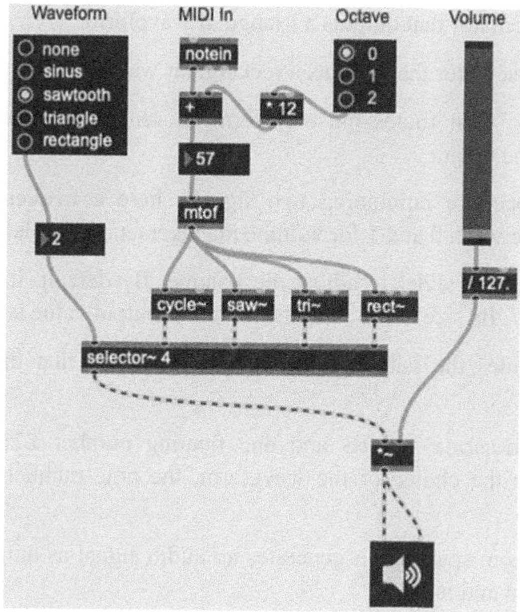

Figure 1.4. *Exercise 1 with Max/MSP*

Instructions and comments are as follows:

– `Waveform`: `radiogroup` object outputting 0, 1, 2, 3, or 4 based on user choice (0 for no signal, 1 for sine signal, 2 for sawtooth signal, 3 for triangle signal, 4 for square wave signal);

– `notein`: MIDI IN acquisition of MIDI notes and information from the keyboard;

– `+; * 12, Octave`: allows you to increase or decrease the notes by +/– 2 octaves by multiplying the output of the vertical object `radiogroup` by 12 and adding it to the note number[3] received, visualized by the object `number` connected to the output of the object +;

– `mtof`: converts the MIDI note number to a frequency in hertz;

– `cycle~`: oscillator that outputs a sine waveform;

– `saw~`: oscillator that outputs a sawtooth waveform;

3 In Appendix 4, you will find the different correspondences of the MIDI note numbers (piano note, frequency, name, etc.).

– `tri~`: an oscillator that outputs a triangular waveform;

– `rect~`: an oscillator that outputs a rectangular waveform;

– `Selector~ 4`: it routes the waveform chosen by the user to the volume control and then the output;

– `*~`: audio operator combining two signals, here a frequency signal and a floating number between 0 and 1 for volume management (`0` by default);

– `Volume`: vertical slider to adjust the volume. By default, it generates values between `0` and 127. Its size is 128, and its minimum output value is 0;

– `/127.`: divides the value of the slider by 127 so that the volume varies between 0 and 1;

– two whole `number` objects and one floating number `flonum` object are present to display the choice of the waveform, the note number and the volume value;

– `Ezdac~` (icon: speaker): it generates an audio signal as output by default on two channels (right and left);

– the texts (waveform, MIDI IN Octave, etc.) are present as comments (`Comment`).

NOTE.– In the event of a malfunction, remember to check the MIDI and audio configuration of Max/MSP in the "Options" menu, "MIDI Setup" sub-menu and "Audio Status". Also, do not forget to click on the "Audio On/Off" icon to the bottom right of the patcher window to activate the sound, and check that the sound is activated on your computer if this is the output you have chosen by default.

1.1.4. *Pure Data*

This exercise covers the main functions in creating a Pure Data patch (see Volume 1, section 4.2.3): object, selector, slider, number, message, etc.

All of these functions will be used throughout the exercises. In Appendix 2, you will find some patches intended to improve or modify the exercises: oscilloscope, keyboard, etc.

NOTE.– You may have seen previously, in section 1.1.3, the term "patcher" for a new project in Max/MSP, whereas here we call it "patch". The terminology is different, but the meaning is the same. It may be an evolution of the language of the developers over time or a problem of rights after the departure of Miller Puckette from the Max/MSP project in 1996 (see Volume 1, sections 4.2.2 and 4.2.3).

The Pure Data exercises in this book have been tested and designed with version 0.53.2; different versions may also be suitable but with some notable differences[4].

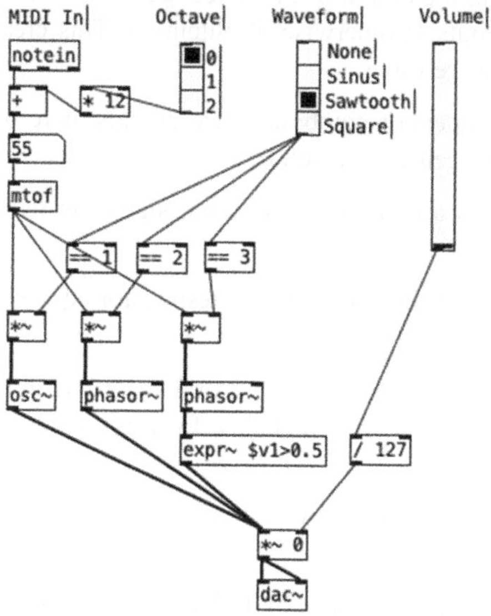

Figure 1.5. *Exercise 1 with Pure Data*

Instructions and comments are as follows:

– `notein`: acquisition of note numbers and MIDI information from the keyboard;

– `+`; `* 12`, `Octave`: it allows you to increase or decrease the notes by +/– 2 octaves by multiplying the output of the vertical object `vradio` by 12 and adding it to the received note number, visualized by the object `number` placed at the output of the object +;

– `mtof`: it converts the MIDI note to a frequency in hertz;

– `*~`: audio operator combining two signals, here the frequency and a binary indicator (0 or 1) validating or not the passage of the signal;

– `osc~`: an oscillator that outputs a sine waveform;

– `phasor~`: an oscillator that outputs a sawtooth waveform;

4 Available at: www.puredata.info/.

– `expr~ $v1>0.5`: a square wave oscillator has a shape that alternates between 0 and 1. Since there is no square wave object in Pure Data, the latter is created by transforming the output of the object `phasor~`. If it is greater than 0.5, the object `expr~` outputs 1; otherwise, it outputs 0. This creates the high (1) and low (0) states of the square wave;

– `*~ 0`: an audio operator combining two signals, here the frequency and an integer for volume management, default 0;

– `dac~`: it generates a default audio output signal on two channels (right and left);

– `== 1, == 2, == 3`: binary operators testing for equality (0 if not equal, 1 if equal);

– `Waveform`: selector outputting 0, 1, 2 or 3 according to the choice of the user (0 for no signal, 1 for a sinusoidal signal, 2 for a sawtooth signal, 3 for a square signal);

– `Volume`: vertical slider to control the volume. By default, its range of values is between 0 and 127 on a linear scale;

– `/127.`: it divides the slider value by 127 to control the volume between 0 and 1;

– Text (MIDI IN Octave, etc.) is present as comments (COMMENT).

NOTE.– In the event of a malfunction, remember to check the MIDI in and audio in configuration in the "Pd" menu, "Preferences" sub-menu, "Audio Settings...", and " MIDI Settings... ". Also, do not forget to check the DSP box or go to the "Media" menu to activate the sound by checking DSP ON, and check that the sound is activated on your computer if this is the output you have chosen by default.

1.1.5. *VCV Rack*

This exercise is designed with the library (fundamental blocks library) provided by default with the VCV Rack. The exercises in this book are for VCV Rack version 2.0, but they may work perfectly on an earlier version, with slight differences.

You will need five modules as follows:

– MIDI-CV: a MIDI input module allowing the connection of peripherals, audio interfaces, keyboard and the selection of the MIDI channel;

– OCT: an octave change module;

– VCO: an oscillator generating four waveforms (sine, triangular, sawtooth and square);

– VCA: a signal amplification module;

– AUDIO: an output module redirecting the signal to the microcomputer or another device such as an audio interface.

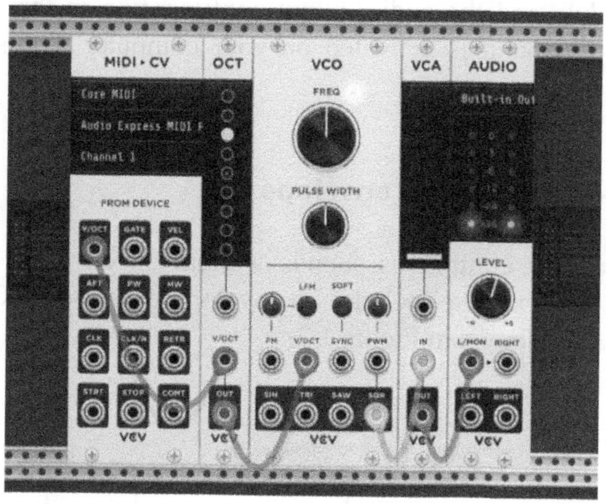

Figure 1.6. *Exercise 1 with VCV Rack. For a color version of this figure, see www.iste.co.uk/reveillac/synthesizers2.zip*

Instructions and comments are as follows:

– place the different modules within the rack;

– connect the V/OCT output of the MIDI-CV module to the V/OCT input of the OCT module;

– connect the OUT output of the OCT module to the V/OCT input of the VCO module;

– connect one of the four outputs, SIN, TRI, SAW, or SQR, of the VCO module to the IN input of the VCA module depending on the waveform you want;

– connect the OUT output of the VCA module to the L/MON input of the AUDIO module. The L/MON input directs the signal to both channels (right and left);

– do not forget to adjust the level of the VCA module (bar graph);

– set the output level, LEVEL of the AUDIO module;

– you can act on the FREQ (frequency) controls if you have chosen SIN (sinusoid), TRI (triangle) or SAW (sawtooth) on the VCO module;

– the PULSE WIDTH control of the VCO module is only active for the SQR output (square).

NOTE.– In the event of a malfunction, remember to check the MIDI configuration in the MIDI-CV module and the audio output configuration in the AUDIO module. Also, check that the sound is activated on your computer if this is your chosen output.

1.2. Exercise 2 – associate an envelope

Exercise 1 provides continuous sounds without an envelope (see Volume 1, section 3.3). This new exercise includes an envelope generator (EG), which will be applied to the VCA, as shown in Figure 1.7.

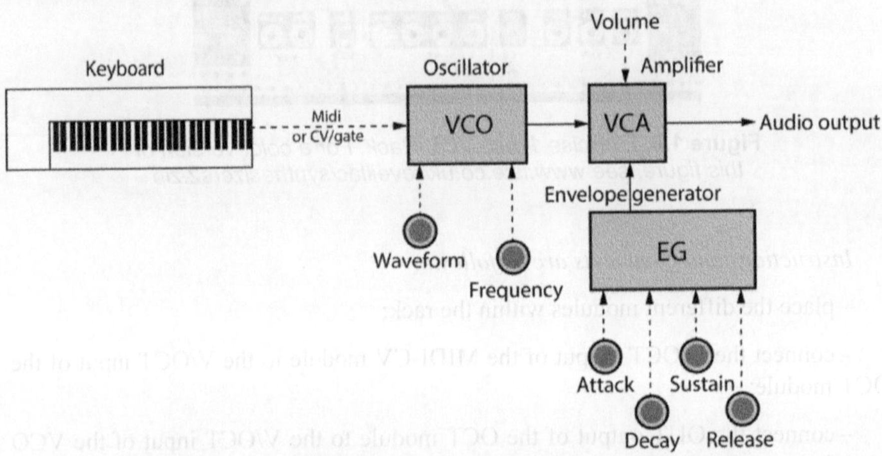

Figure 1.7. *Subtractive sound synthesis with an oscillator, an envelope generator and an amplifier (VCO – EG – VCA). The command signals are shown with dotted lines*

You will find the classic audio signal shaping parameters: attack, decay, sustain, and release (ADSR). The input controller remains a keyboard.

1.2.1. *Behringer Neutron*

We keep the patch cable between OSC1 OUT and VCA IN for this second exercise on the Neutron matrix. A pre-routing of the EG on the VCA already exists by default.

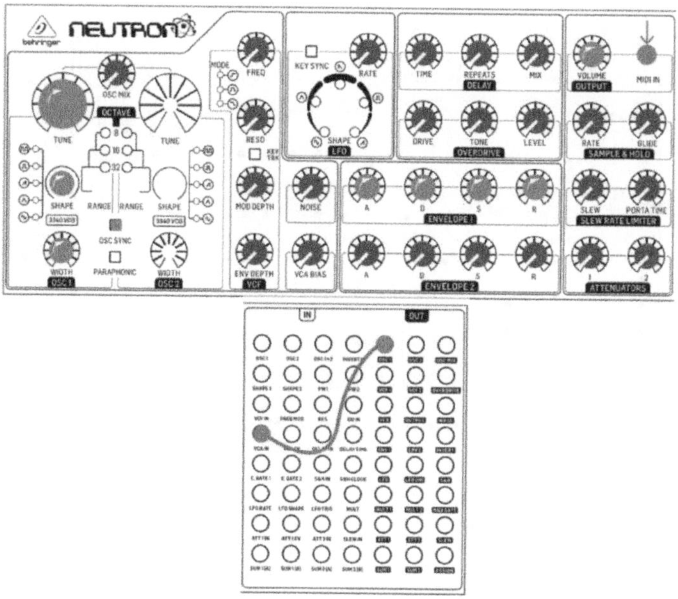

Figure 1.8. *Exercise 2 for the Behringer Neutron. For a color version of this figure, see www.iste.co.uk/reveillac/synthesizers2.zip*

Instructions and comments are as follows:

– keep the configuration in exercise 1;

– as before, the controls for oscillator 1 (OSC1) are all accessible, as well as the general volume knob;

– the VC BIAS knob (VCA influence) must be turned all the way to the left to close the VCA to the maximum; otherwise, the signal coming from the oscillator will be diverted and the EG will have no effect;

– the VOLUME OUTPUT knob can be adjusted to adjust the overall audio output volume to your convenience;

– the A, D, S, R buttons allow you to adjust the envelope parameters. Note that it is envelope 1 that acts on the VCA (see Volume 1, section 4.1.3, Figure 4.24).

1.2.2. *Behringer 2600 (ARP 2600)*

To set up the EG acting on the VCA, we will modify the configuration adopted in exercise 1.

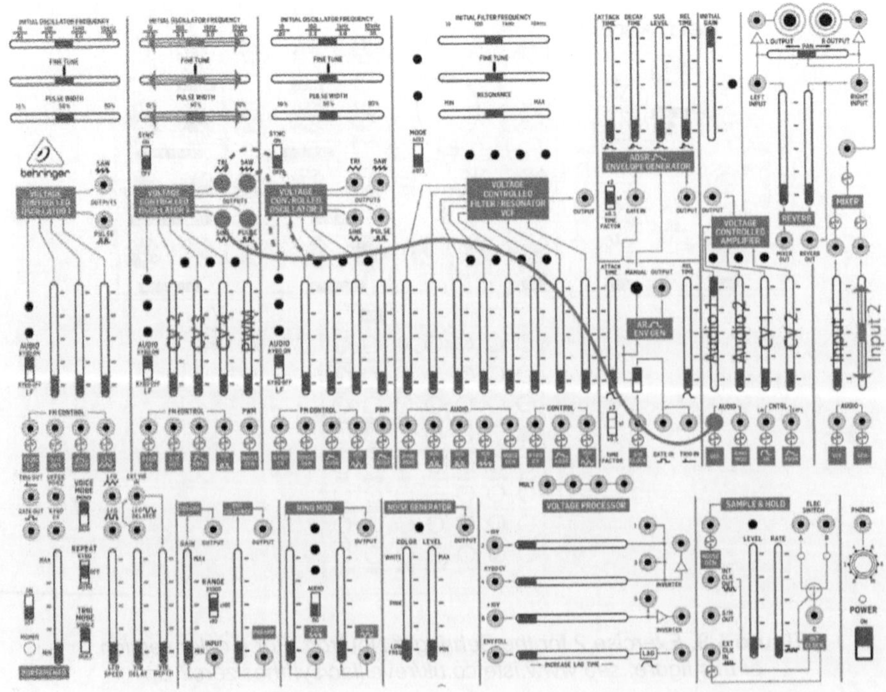

Figure 1.9. *Exercise 2 for Behringer 2600 (ARP 2600). For a color version of this figure, see www.iste.co.uk/reveillac/synthesizers2.zip*

Instructions and comments are as follows:

– keep the wiring from the configuration in exercise 1;

– set the INITIAL GAIN control to the down position to completely reduce the overall gain of the VCA;

– the audio signal input control CV2 must be placed at its maximum (highest position of the slider);

– the ATTACK TIME, DECAY TIME, SUS LEVEL, REAL TIME (ADSR) controls of the EG can be manipulated to define the envelope curve of the input signal;

– the audio input 1 control of the MIXER module controls the output volume.

1.2.3. *Max/MSP*

Exercise 2 will enrich the patcher already created in exercise 1 with new functionalities.

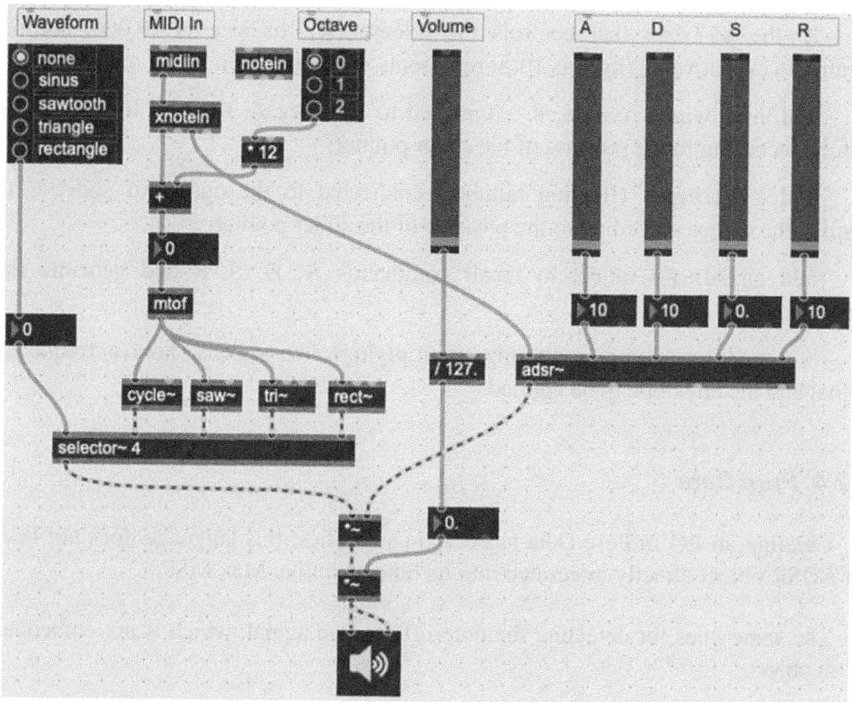

Figure 1.10. *Exercise 2 with Max/MSP*

Instructions and comments are as follows:

– `midiin`: replace the object `notein` to capture the notes coming from the keyboard;

– `xnotein`: connected to the `midiin` object. The `xnotein` provides an additional logical indicator (0 or 1) with respect to `notein`. It detects if a key has been pressed on the keyboard. This indicator will be connected to the input of `ADSR~` to signal to it the state of the keyboard keys, pressed or released (MIDI note-on, note-off);

– `A`: the `slider` that controls the attack (A). Its size is 10,000, and its output minimum (min) value is 10. It represents a time between 10 ms and 10 s;

– D: the `slider` that controls the decay (D). Its size is 1,000, and its minimum output value (min) is 10. It represents a time between 10 ms and 1 s;

– S: the `slider` that controls the sustain (S). Its size is 1, and its minimum output decimal value (min) is 0. It represents a percentage between 0 and 10%;

– R: the `slider` that controls the release (R). Its size is 10,000, and its minimum output value (min) is 10. It represents a time between 10 ms and 10 s;

– add three whole `numbers` connected to `slider` A, D and R to display the values according to the position of the slider pointer;

– add a `flonum` (floating number) connected to the output of slider `S` to display the values according to the position of the slider pointer;

– add an `ADSR~` object to recall commands A, D, S, R and generate the envelope;

– `*~`: audio operator combining (multiplying) two signals, here a frequency signal and the envelope to be applied.

1.2.4. *Pure Data*

Creating an EG in Pure Data has certain subtleties; this language does not have an ADSR object directly integrated into its library, unlike Max/MSP.

The same goes for detecting the note-off/note-on signal, which is not embedded in an object.

Instructions and comments are as follows:

– A: `slider` that controls the attack (A). Its linear range is between 10 and 10,000. It represents a time in milliseconds;

– D: `slider` that controls the decay time (D). Its linear range is between 10 and 1,000. It represents a time in milliseconds;

– S: `slider` that controls the sustain level (S). Its linear range is between 0 and 1. It represents percentages (0–100%);

– R: `slider` that controls the release time (R). Its linear range is between 10 and 10,000. It represents a time in milliseconds;

– the four `number` objects present behind the `slider` objects display the values of each of them;

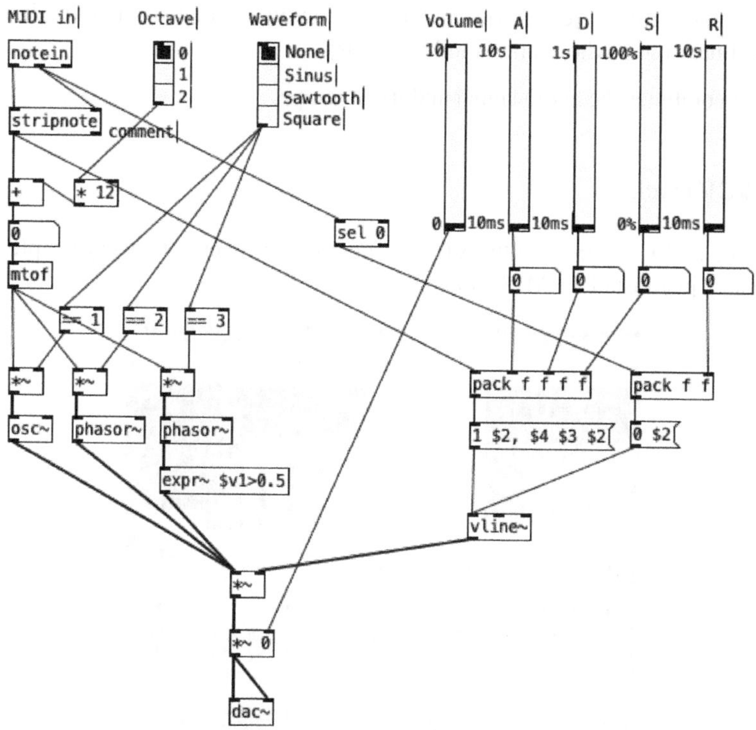

Figure 1.11. *Exercise 2 with Pure Data*

– `Stripnote`: detects the presence of a pressed key (note-on) to trigger the `pack` object, which will define the attack, decay and sustain (A, D and S);

– `Sel`: makes it possible to test if its input, the velocity, is equal to the constant 0. If so, the left output is true (1); otherwise, the right output gives the velocity. When velocity disappears, release can take place;

– `Pack` (on the left): concatenates all the entries, that is, they are placed one after the other. The input data are the note number, attack, decay and sustain (A, D and S). As the output, `pack` will provide the value of 1, that is, an upward ramp, necessary for the attack, followed by the three arguments;

– `Pack` (right): if its right input is true (1), that is, if the key has been released (zero velocity), then the output is equal to 0 followed by the release (R). The value 0 indicates a downward ramp, which is required for a release;

– behind the two `pack` objects place a `message` object, which retrieves the sequence of values to be transmitted to the `vline~` object;

– `vline~`: this object retrieves the data on its input to transform it into a variable linear ramp whose unit is milliseconds;

– `*~`: combines the audio signal and its envelope.

1.2.5. *VCV Rack*

Adding an EG controlling the VCA to the previous exercise is very simple. With VCV Rack, just install an ADSR module.

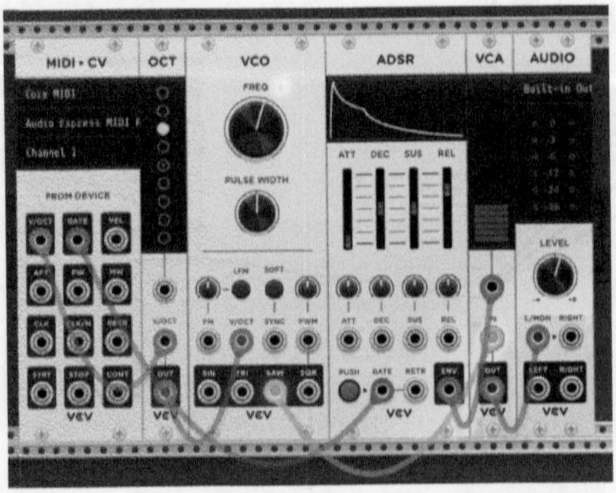

Figure 1.12. *Exercise 2 with VCV Rack. For a color version of this figure, see www.iste.co.uk/reveillac/synthesizers2.zip*

Instructions and comments are as follows:

– place an ADSR module between the VCO and the VCA;

– connect the GATE output of the MIDI-CV module to the GATE input of the ADSR module;

– connect the ENV output of the ADSR module to the control input of the VCA;

– by acting on the four commands of the ADSR module, you can vary the contours of the output signal.

1.3. Exercise 3 – add a filter

To continue, we will enrich our work with a filter, which will process the signals from the VCO. We will choose a low-pass filter with cutoff frequency and resonance management, as shown in Figure 1.13.

Other filters are possible (high-pass, band-pass, with different types and more complex settings). They are often present on synthesizers, with each manufacturer trying to offer more sound processing possibilities.

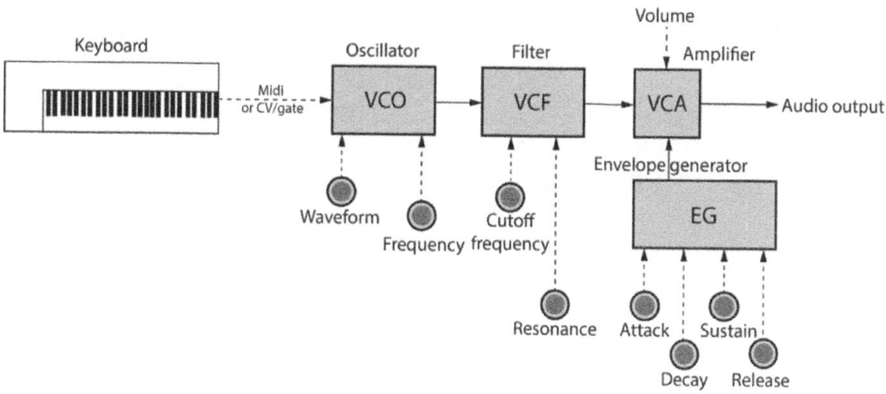

Figure 1.13. *Subtractive sound synthesis with an oscillator, filter, envelope generator and amplifier (VCO – VCF – EG – VCA). The command signals are shown with dotted lines*

1.3.1. *Behringer Neutron*

Compared to the previous exercise, placing one or more patch cables on the Neutron matrix is unnecessary. By default, a pre-routing exists between oscillator 1 (OSC 1) and the filter (VCF).

Instructions and comments are as follows:

– remove the patch cable from exercise 2;

– turn the LEVEL knob of the OVERDRIVE entirely to the right to authorize the output of the audio signal to the VCA;

– the three buttons of the VCF allow you to adjust the cutoff frequency of the filter (FREQ), its resonance (RESO) and the modulation depth (MOD DEPTH);

– the type of filter can also be changed by pressing the MODE button (from top to bottom: high-pass filter, band-pass filter, low-pass filter);

– the setting of envelope 1 (ADSR) relating to the VCA always remains active, like the settings of oscillator 1 and the control of the general volume;

– do not forget to place the VCA BIAS button at its minimum to preserve the action of the EG 1.

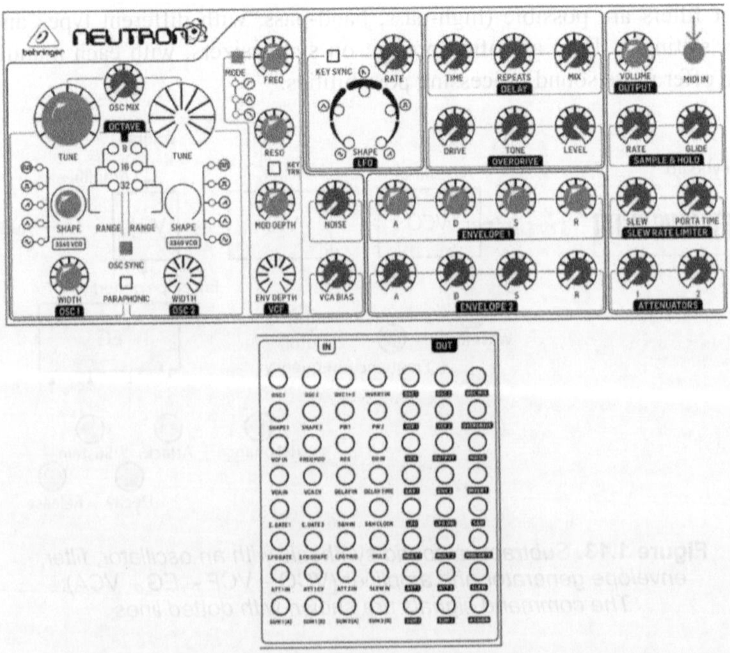

Figure 1.14. *Exercise 3 for Behringer Neutron. For a color version of this figure, see www.iste.co.uk/reveillac/synthesizers2.zip*

1.3.2. *Max/MSP*

For this exercise, we are going to add a low-pass resonant filter, whose cutoff frequency and resonance we can adjust by integrating an object into our patcher `lores~`.

Instructions and comments are as follows:

– repeat the patcher from exercise 2;

– `Cutoff frequency`: `slider` controlling the cutoff frequency of the low-pass filter. It has a size of 50 and a minimum value of 0;

– `Resonance`: `slider` controlling the resonance of the low-pass filter. It has a size of 1 and a minimum decimal value of 0;

– add a whole `number` to display the value of the `slider Cutoff frequency` object;

– add a floating number object `flonum` to display the value of the `slider Resonance` object;

– `lowres~`: resonant low-pass filter, the right input picks up the frequency signal, the center input picks up the cutoff frequency and the left input picks up the resonance.

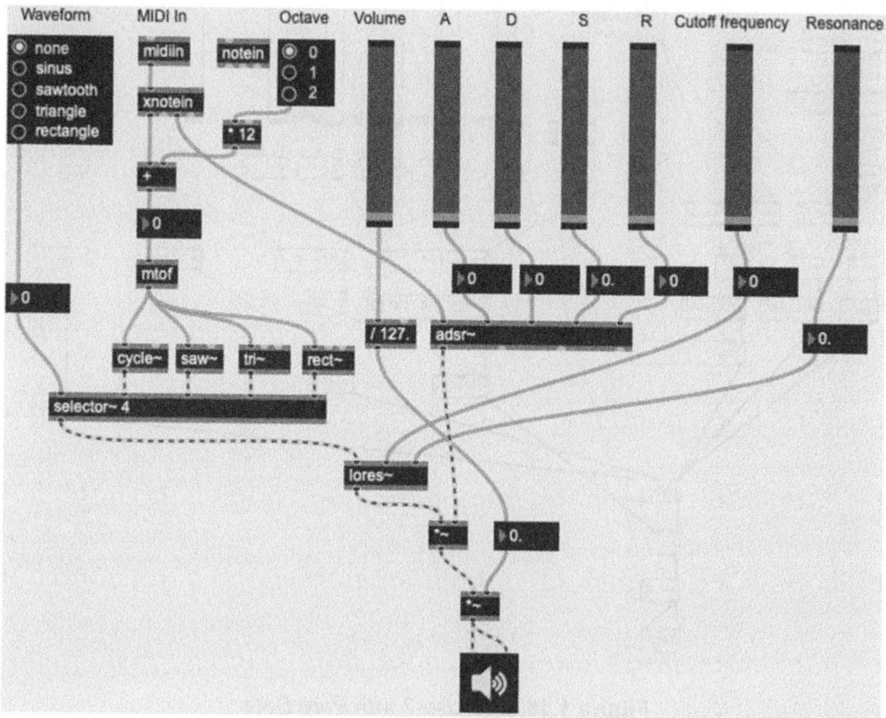

Figure 1.15. *Exercise 3 with Max/MSP*

1.3.3. *Pure Data*

There are several objects that can act as a filter. In this exercise, we will use the `bob~` filter (this name pays homage to Robert Moog, one of the inventors of the first analog synthesizer).

Its characteristics make it a filter well suited to subtractive synthesis, it is a Runge-Kutta type[5] 4 (RK4) filter and it perfectly simulates the resonant filter used on Moog synthesizers.

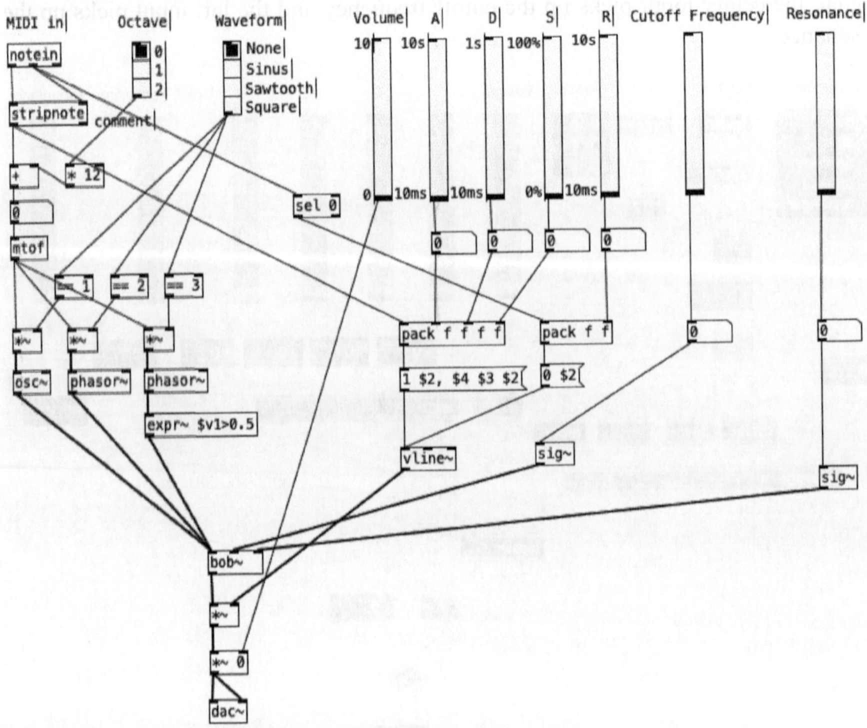

Figure 1.16. *Exercise 3 with Pure Data*

Instructions and comments are as follows:

– resume the setup of the previous exercise;

– `Cutoff frequency`: `slider` that sets the filter cutoff frequency. Its logarithmic range is between 0 and 10,000;

– `Resonance`: `slider` that controls the filter cutoff frequency. Its linear range is between 0 and 6;

5 Numerical approximation method developed in 1901 by Carl Runge and Martin Wilhelm Kutta, widely used, among other things, in filter design calculations.

– `sig~`: an object that converts a number into an audio signal;

– `bob~`: filter whose three inputs, from left to right, are the frequency signal, the filter cutoff frequency and the resonance. Beyond 4, the filter goes into self-oscillation;

– both `number` objects display the values of each of the `Cutoff frequency slider` and `Resonance slider` objects.

1.3.4. *VCV Rack*

Let us continue to evolve our virtual modular synthesizer by adding a VCF module.

Figure 1.17. *Exercise 3 with VCV Rack. For a color version of this figure, see www.iste.co.uk/reveillac/synthesizers2.zip*

Instructions and comments are as follows:

– repeat exercise 2;

– remove the patch cable between the VCO and the VCA;

– add a VCF module;

– connect one of the VCO outputs (SIN, TRI, SAW or SQR) to the input (IN) of the VCF;

– connect the HPF (high-pass frequency) output of the VCF to the IN input of the VCA. You can also test the LPF (low-pass frequency) output to listen to the influence of each filter on the sound reproduction.

1.4. A little further with Max/MSP, Pure Data and VCV Rack

Below are additional elements for creating envelopes and filters with virtual synthesis software, Max/MSP and Pure Data.

A section will also be devoted to the VCV Rack to introduce us to some modules that differ from those provided by default (fundamental blocks library).

We carefully study these different patches, which we can integrate into previous exercises.

1.4.1. *ADSR with Max/MSP*

Here is another simpler way (no `midiin` or `xnotein` object) to generate an ADSR envelope using the velocity of a `notein` object to check the note-off/note-on triggering of the keyboard.

By default, the chosen oscillator produces a rectangular signal, but it is easy to adapt this example.

Instructions and comments are as follows:

– object`/127.`: it divides the velocity, which varies between 0 and 127, to generate, at the output, a decimal value between 0 and 1. When it is not 0, a key on the keyboard is detected as pressed, and the release is not considered;

– the `flonum` object placed behind the object`/127.` displays the decimal value of the division (between 0 and 1);

– the four `dial` objects are rotary cursors that vary each of the parameters A, D, S and R. Their sizes are 2001, 1001, 1 (decimal) and 2001. Their minimum values are all equal to 0;

– to accurately control the parameter values, three `number` objects for A, D, R and one `flonum` object for S connect to the `dials` located above;

– the object `rect~` already encountered in the exercises generates a rectangular signal.

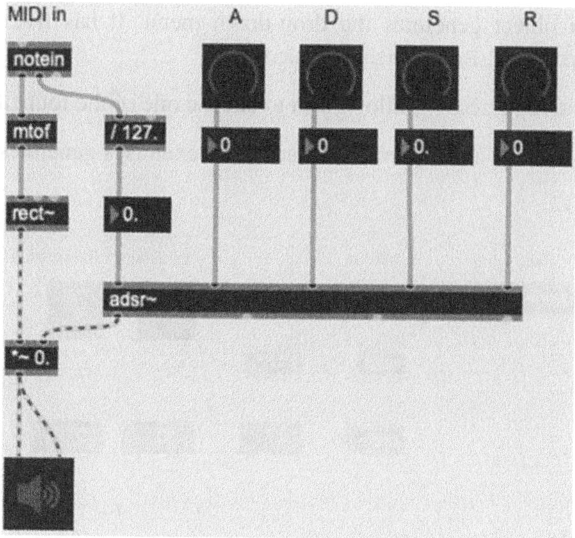

Figure 1.18. *Easy ADSR management*

1.4.2. *Filter with Max/MSP*

For this example, we use an `svf~` object, a multi-state filter that produces four outputs: low-pass, high-pass, band-pass or notch.

Instructions and comments are as follows:

– the object`/127.`: it divides the velocity, which varies between 0 and 127, to generate, at the output, a decimal value between 0 and 1. When it is not equal to 0, a key on the keyboard is detected as pressed, and the release is not considered;

– the `flonum` object, which is placed behind the object`/127`, displays the decimal value of the division (between 0 and 1);

– the two `dial` objects are rotary sliders that vary the filter cutoff frequency and resonance. Their sizes are, respectively, 20,000 and 1 (decimal). Their minimum values (min) are equal to 0;

– to manage parameter values more precisely, two `number` objects for the cutoff frequency and the resonance connect to the dials located above;

– the object `svf~` is the filter; it has four outputs corresponding to each filter type. From left to right, the order is low-pass, high-pass, band-pass and rejector;

– an umenu object generates the drop-down menu. It has five items: None, Lowpass, Highpass, Bandpass and Notch;

– the object selector~ 4 allows you to choose one of the four filter outputs;

– the object rect~ already encountered in the exercises generates a rectangular signal.

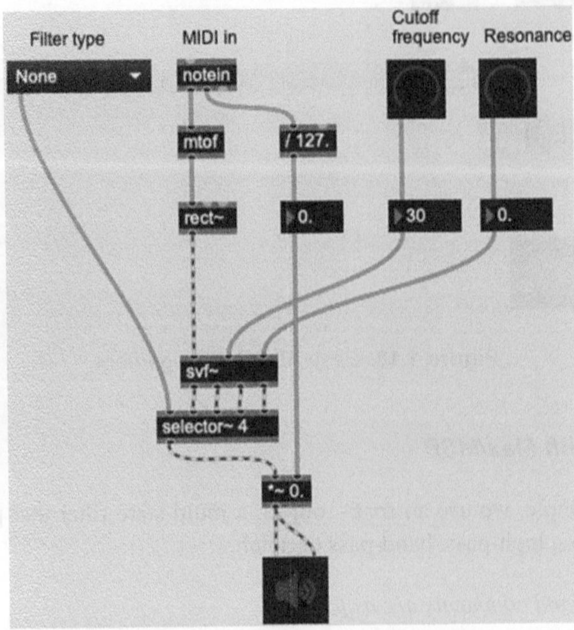

Figure 1.19. *Using a multistate filter*

1.4.3. *Configurable filter with Max/MSP*

This example uses a powerful feature of the Max/MSP, the filtergraph~ object, which provides a graphical interface allowing you to fully configure a biquad~ or cascade~ filter.

Here, we will use the biquad~ (biquadratic) object, a linear recursive second-order filter with two poles and two zeros.

A spectroscope~ object displays the spectrogram of the signal without filtering. This object can also display a sonogram depending on its configuration.

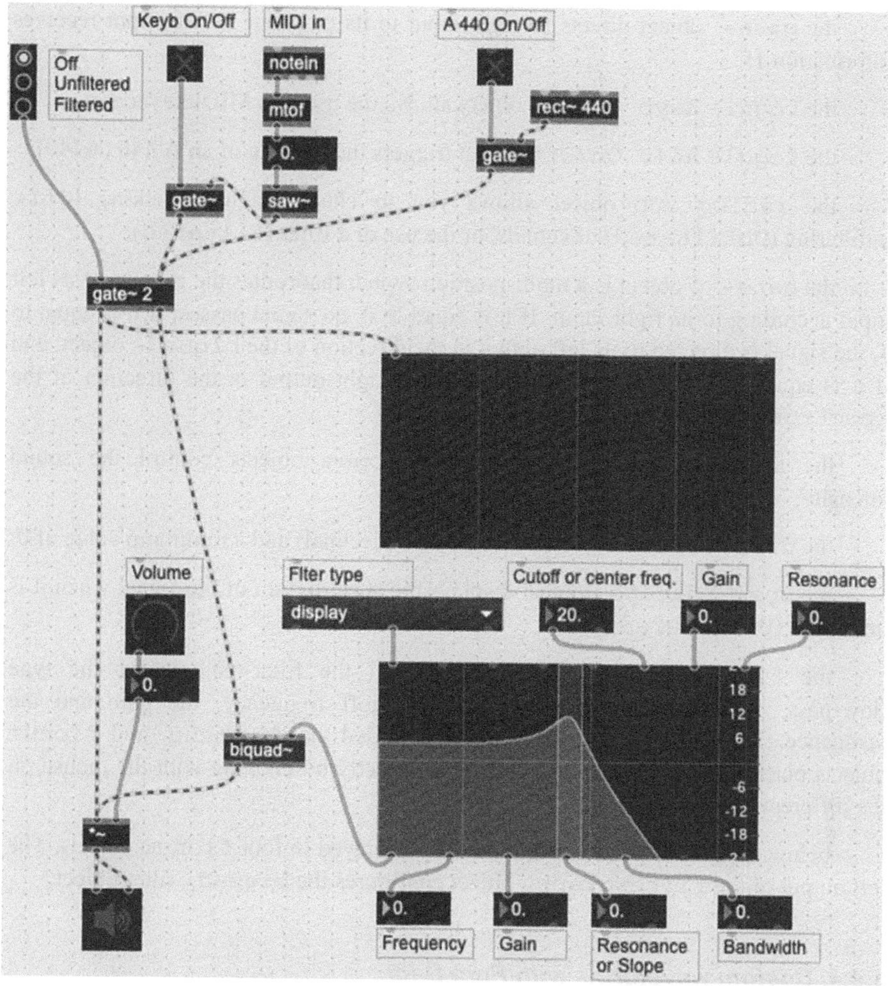

Figure 1.20. *Use of a configurable filter*

Instructions and comments are as follows:

– the `rect~` object generates a square wave signal;

– `Keyb On/Off` and `A440 On/Off` represent two `toggle` objects, that is, on/off switches (0 or 1), which each control a `gate~` object;

– the gate~ object directs its right input to its output if its left input receives information 1;

– the toggle Keyb On/Off object allows the use of a MIDI keyboard;

– the toggle A440 On/Off object triggers the sending of an A 440 (A440);

– the radiogroup object allows you to choose signal blocking (Off), unfiltering (Unfiltered) of signals, or the use of a filter (Filtered);

– the gate~ 2 object is a multi-position switch that routes the content of its left input according to its right input. If it is equal to 0, no signal passes; if it is equal to 1, the signal is directed to its left output in the direction of the biquad~ object; and if it is equal to 2, the signal is directed to its right output in the direction of the spectroscope~ object;

– the associated dial Volume and flonum objects control the sound intensity;

– the dial Volume object has a size of 1 (decimal) and a minimum value of 0;

– the spectroscope~ object displays the spectrogram of the signal when it is unfiltered (Unfiltered);

– the filtergraph~ object manages all the filter parameters, the type (low-pass, high-pass, band-pass, etc.), the cutoff frequency, the gain and the resonance. These four parameters can be modified via the umenu and flonum objects placed above the display window or directly by clicking with the mouse on the different parts of the graph;

– below the graph, the changes made are displayed in four flonum objects. The left output of the filtergraph~ object configures the biquad~ filter object.

1.4.4. *Custom waveforms with Pure Data*

As you may have noticed in the previous exercises, Pure Data can generate classic, sinusoidal waveforms (osc~), sawtooth (phasor~) or square (see section 1.1.4).

It is possible to define a custom waveform by acting on the filtering of the audio signal, as subtractive synthesis skillfully does; however, there is another way, using waveshaping, as shown in Figure 1.21.

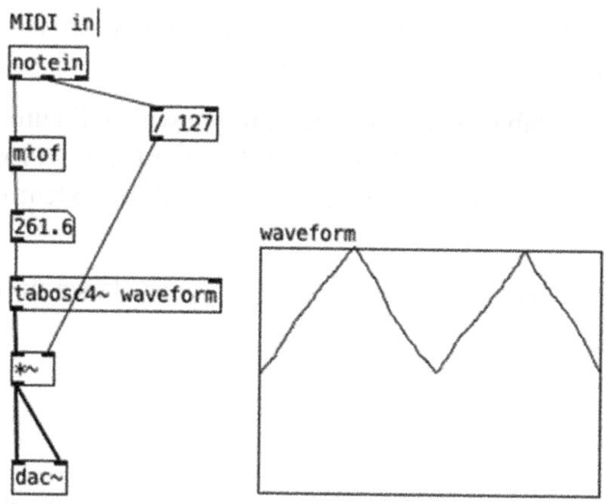

Figure 1.21. *Custom waveform*

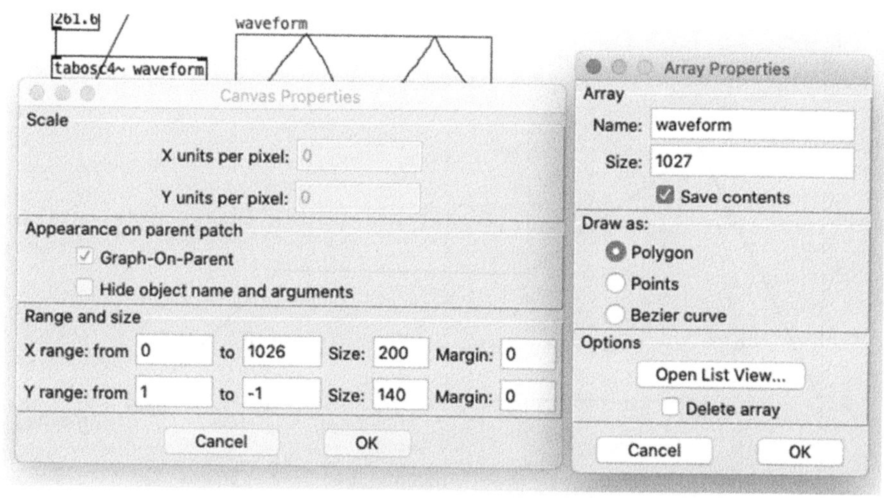

Figure 1.22. *Waveform array parameters. For a color version of this figure, see www.iste.co.uk/reveillac/synthesizers2.zip*

Instructions and comments are as follows:

– the `notein`, `mtof`, `number`, `/127`, `*~` and `dac~` objects have already been encountered in previous exercises. We will not go back over their use;

– `tabosc4~` applies to the `waveform` table. It is a wavetable oscillator operating on a 4-point polynomial interpolation. The number of values (points of the plot) in the table must equal a power of 2 to which 3 points are added. For example, for 29, we will have 512 + 3 = 515 values and for 2^{10}, 1,024 + 3 = 1,027 values;

– `waveform` is a table whose settings are shown in Figure 1.22. For Array Properties, the name is waveform, the size is 1,027. For Canvas Properties, the range of X varies between 0 to 1,026, or 1,027 values, and the extent of Y varies from 1 to –1;

– with your mouse, you can draw the curve of your choice in the frame defined by the `waveform` object.

1.4.5. *ADSR with Pure Data*

In exercise 2 (see section 1.2.4), we set up an ADSR (attack, decay, sustain, release) EG. In this example, we will use another solution based on delays to shape the audio signal, while keeping the same parameters.

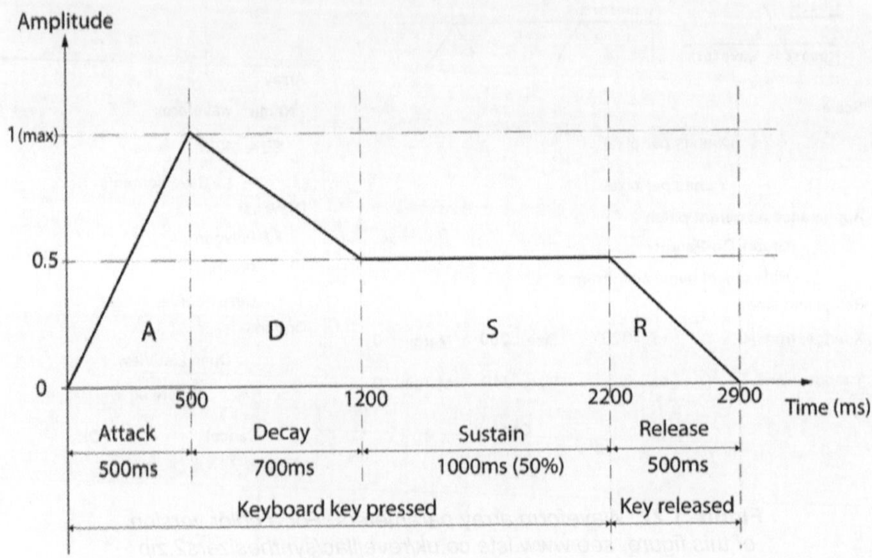

Figure 1.23. *ADSR curve*

The principle is built around the succession of different time ranges linked by `delay` objects (abbreviated `del`).

Take the example of Figure 1.23; the selected ADSR curve respects the following parameters:

– time: A = 500 ms, D = 700 ms, S = 1,000 ms and R = 500 ms;

– intensities: A varies from 0% to 100% (1), D varies from 100% to 50% (0.5), S remains at 50% (0.5), and R varies from 50% to 0% (0).

In our Pure Data patch, we will implement different delays and intensities.

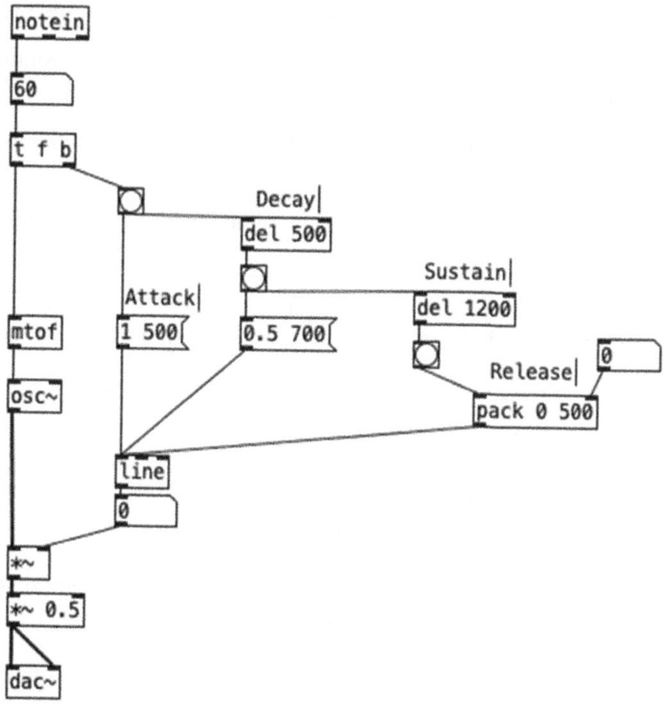

Figure 1.24. *ADSR and delays*

Instructions and comments are as follows:

– the objects `notein`, `number`, `mtof`, `osc~`, `*~`, `*~ 0.5` and `dac~` have already been seen in the exercises;

– the three `bang` objects are there to indicate the sequence of the different stages that the ADSR curve follows;

– the `message` object (`Attack`) transmits an intensity ramp equal to 1 (100%) and of 500 ms duration;

– the `del` object (`Decay`) sets a wait of 500 ms corresponding to the duration of the attack;

– the `message` object (`Decay`) transmits an intensity ramp of 0.5 (50%) and a duration of 700 ms;

– the `del` object (`Sustain`) defines a wait of 1200 ms, which corresponds to the sum of the duration of the attack (A: 500 ms) and the decay (D: 700 ms);

– the `pack` (`Release`) object defines a release (R) to be transmitted with a duration of 500 ms;

– the `number` object connected to the right `pack` input allows the user to vary the duration of the release (R);

– the `line` object generates the different ramps, which are multiplied by the audio signal;

– the `number` object placed at the output of the `line` object shows the numerical sequence of the different temporal steps (ramps) of the ADSR curve.

1.4.6. *Signals and filters with Pure Data*

We have seen that filters are vital parts of subtractive sound synthesis; all synthesizers have at least one, with advanced settings.

The following example will allow the user to discover the main types of filters present within Pure Data: `vcf~`, `lop~`, `hip~`. and `bp~`.

To feed these filters, the user will have two types of waveforms at their disposal: triangular or symmetrical sawtooth.

Instructions and comments are as follows:

– only the Pure Data objects necessary to understand this example will be presented;

– for the oscillator generating triangular signals, we use four objects: `phasor~` (generation of the sawtooth signal), `clip~ 0 0.5` (signal truncation), `*~ 2` (signal amplification by 2) and `-~` (signal subtraction);

– Figure 1.26 shows the progression of the waveform through these four objects;

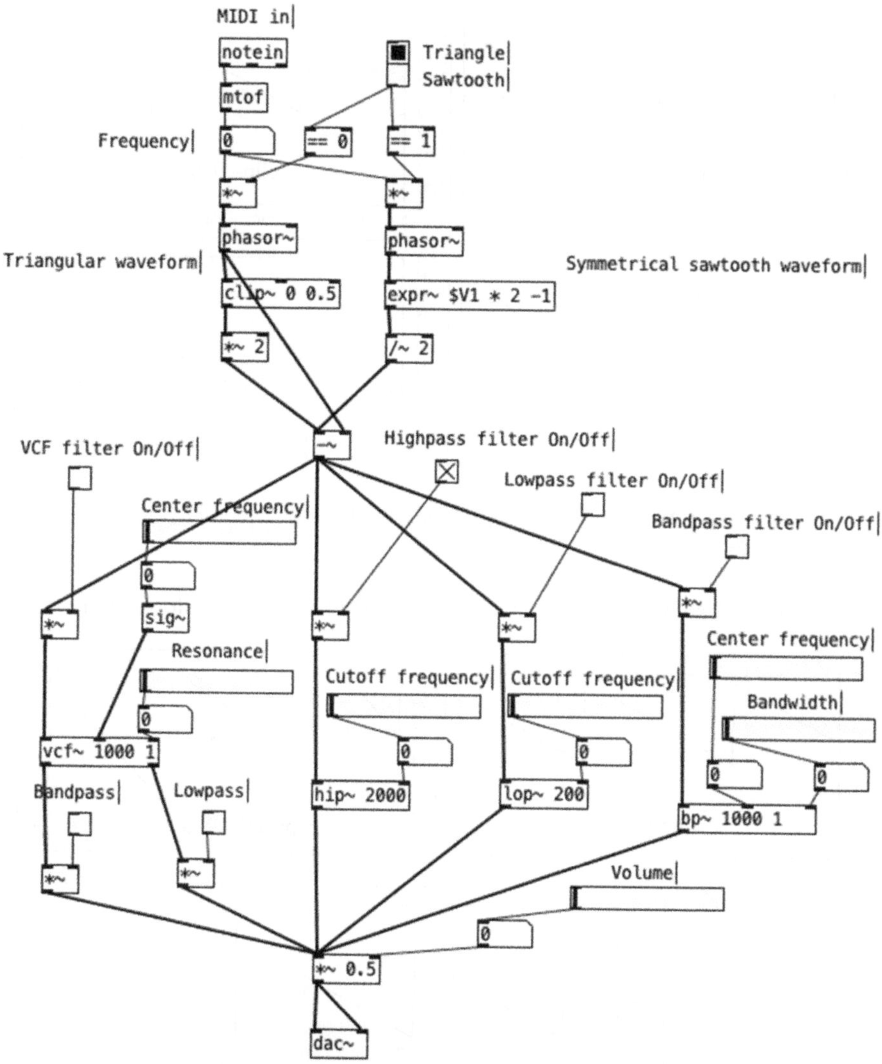

Figure 1.25. *Signals and filters*

– for the oscillator generating sawtooth signals, we use three objects: `phasor~` (generation of the sawtooth signal), `expr~ $V1 * 2 - 1` (transformation of a unipolar signal into a bipolar signal) and `/~ 2` (signal attenuation by 2);

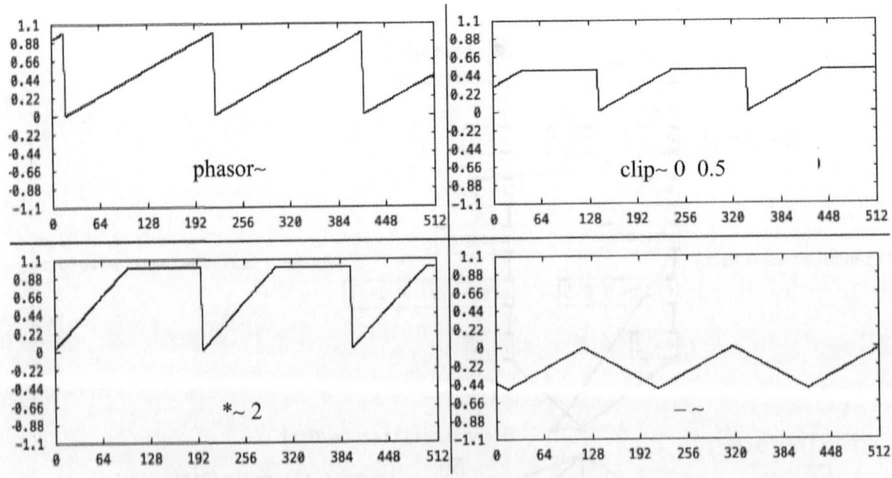

Figure 1.26. *Successive waveforms to obtain a triangular signal*

– Figure 1.27 shows the successive forms of the waveform transformation;

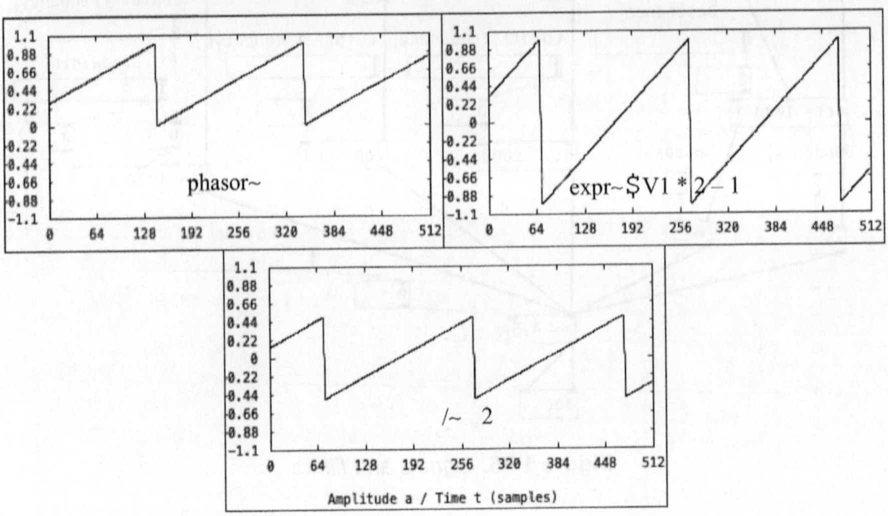

Figure 1.27. *Successive waveforms for the sawtooth signal*

– the four `slider Cutoff frequency` and `Center frequency` objects have a logarithmic range of values between 10 and 10,000;

– the `Resonance slider` object has a range of linear values between 0 and 20;

– the `Bandwidth slider` object has a range of linear values between 0 and 10;

– the `volume slider` object has a range of linear values between 0 and 1;

– the four objects `toggle`, `VCF Filter On/Off`, `Highpass filter On/Off`, `Lowpass filter On/Off` and `Bandpass filter On/Off` allow you to select the filter you want to activate. They are cumulative;

– the two `toggle` objects, `Bandpass` and `Lowpass`, allow you to choose the type of filter to use for the `VCF~` object;

– under each `slider` object, a `number` object indicates the chosen value;

– by default, each filter has pre-existing parameters, for example, a cutoff frequency of 1,000 Hz and a resonance equal to 1 for the `VCF~` object;

– do not hesitate to try all the possible settings; you will be able to better understand the use and the modifications made to the sound reproduction by each of the filters on waveforms with different harmonics, odd for the triangular signal and even/odd for the sawtooth signal;

– for the curves corresponding to each signal, see Appendix 2, section A2.1, to integrate a control oscilloscope patch.

1.4.7. *Additional modules with the VCV Rack*

The VCV Rack has a vast library of modules, many of which are virtual copies of actual off-the-shelf hardware modules.

Many of these software modules are free and often have much more advanced functions than those provided in the fundamental blocks library.

In Figure 1.28, you can see a configuration showing an alternative to exercise 3 using the VCO module "Macro Oscillator" from Audible Instruments (free) and the VCF module "Stabile" from Vult (free).

Instructions and comments are as follows:

– wire the two modules;

– try the different settings on each of the modules;

– for the Macro Oscillator, edit the different waveforms by turning the EDIT button;

– for Stabile, you can use each of the outputs (LP: low-pass, BP: band-pass, HP: high-pass and SEM: semblance).

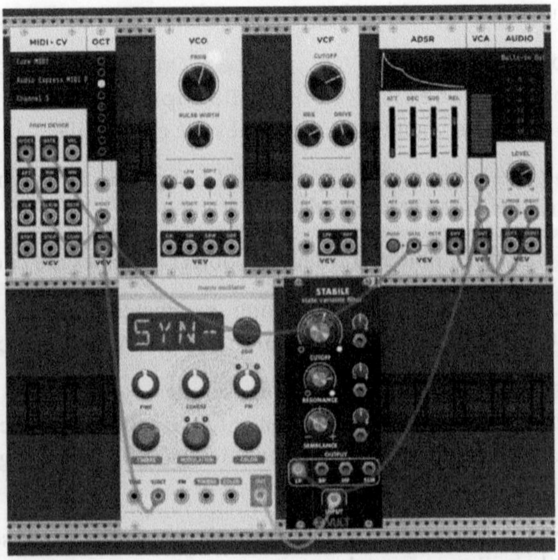

Figure 1.28. *Macro Oscillator and Stabile in a VCV Rack. For a color version of this figure, see www.iste.co.uk/reveillac/synthesizers2.zip*

1.5. Final remarks

In this chapter, we have seen the beginnings of subtractive synthesis on different pieces of hardware and software, but we are still far from having synthesizer models worthy of being called such.

The various exercises presented provide solutions but they are not the only ones, particularly with regard to the VCV Rack, Max/MSP and Pure Data, which are more open than the Behringer Neutron or the 2600.

It is interesting to find other configurations that can provide answers to the three exercises proposed. Consult the reference section at the end of the book to deepen your knowledge. Do not hesitate to ask questions on the many specialized forums in the Internet links section.

2

Subtractive Synthesis: The Fundamentals

After an introductory chapter to master the basics of subtractive sound synthesis, it is time to approach other techniques that make a synthesizer a genuine instrument of research and sound creation.

Adding a dedicated filter envelope, a low-frequency oscillator (LFO), a noise generator, and several oscillators will provide the thickness necessary for creating rich and complex signals.

Features like ring modulation or sample and hold will further complement the subtractive sound synthesis toolbox.

Finally, the various effects will provide the finishing touch, producing an even more spectacular final sound rendering.

In this chapter, I will cover exercises with hardware synthesizers (Minimoog, Novation Bass Station II, ARP 2600, Arturia MatrixBrute) while retaining my software tools (Max/MSP, Pure Data, VCV Rack) to which I will add Native Instruments Reactor.

You will find in Table 2.1, for each of the exercises, the machines and software chosen, and the sections where you can find them.

Exercise	Hardware/software	Section
No. 4. 1 VCO, 1 VCA, 2 GE	Minimoog (or clones)	2.1.1
	Behringer 2600 (ARP2600)	2.1.2
	Max/MSP	2.1.3
	Pure Data	2.1.4
	VCV Rack	2.1.5
No. 5. 1 VCO, 1 VCA, 2 GE, 1 LFO	Minimoog (or clones)	2.2.1
	Novation Bass Station II	2.2.2
	Max/MSP	2.2.3
	Pure Data	2.2.4
	VCV Rack	2.2.5
	Reaktor	2.2.6
No. 6. Various VCO	Minimoog (or clones)	2.3.1
	Novation Bass Station II	2.3.2
	Max/MSP	2.3.3
	Pure data	2.3.4
	VCV Rack	2.3.5
	Reaktor	2.3.6
No. 7. Noise generator	Minimoog (or clones)	2.4.1
	Behringer 2600 (ARP2600)	2.4.2
	Novation Bass Station II	2.4.3
	Max/MSP	2.4.4
	Pure Data	2.4.5

Table 2.1. *Summary of the different exercises*

2.1. Exercise 4 – adding an envelope to the filter

In Chapter 1, we saw that the envelope generator can be applied to the VCA to control its amplitude according to the four ADSR (attack, decay, sustain, release) parameters.

The envelope is an amplitude curve that gives the dynamics of the sound over time, with that of a percussive instrument being very different from that of a flute, the backwash of waves, or a storm.

Here, we will focus on the envelope applied to the filter to manipulate its opening and closing. Depending on its type, it acts on the time curve to control its cutoff frequency, resonance or other parameters.

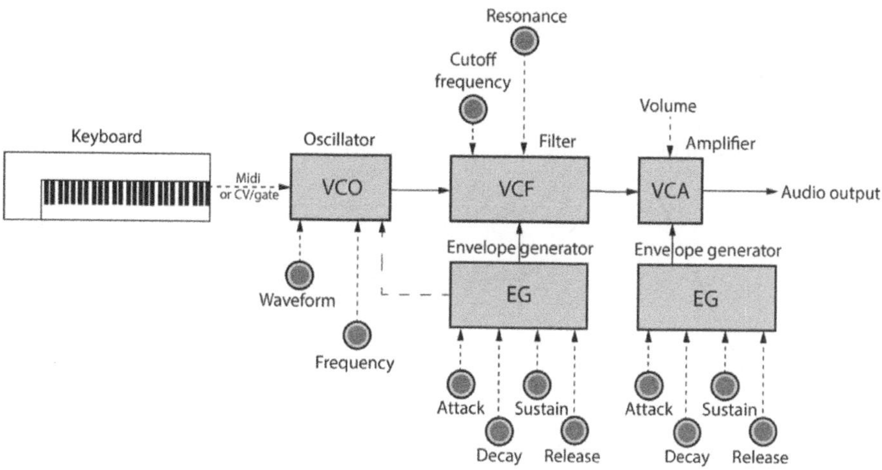

Figure 2.1. *Subtractive sound synthesis with an oscillator, two envelope generators and an amplifier (VCO – EG – VCA). The command signals are in fine dotted lines*

NOTE.– The envelope generator can also be applied in some cases to the VCO to add a detuning movement (change in pitch) to the attack of the sound, often used with percussion instruments (in reality, when a stick or hand hits a tense skin like that of a tom-tom or a conga drum, a slight detuning is created due to the distension during impact).

2.1.1. *Minimoog*

The Minimoog is a reference instrument. It is a monodic synthesizer with the most classic architecture with two envelope generators, one for the VCF and one for the VCA.

It should be noted that unlike many machines, the envelopes of the Minimoog are based on only three parameters: attack, decay and sustain.

The release found on many other hardware is absent. To overcome this, there are two workarounds, the double use of decay and a switch allowing the release to be reduced to 10 ms.

Figure 2.2 gives an overview of the envelope curves controlled by the Minimoog. The dotted part of the curve corresponds to the setting (On) of the decay switch located above the Pitch and Mod (Modulation) wheels. In this case, the release is limited.

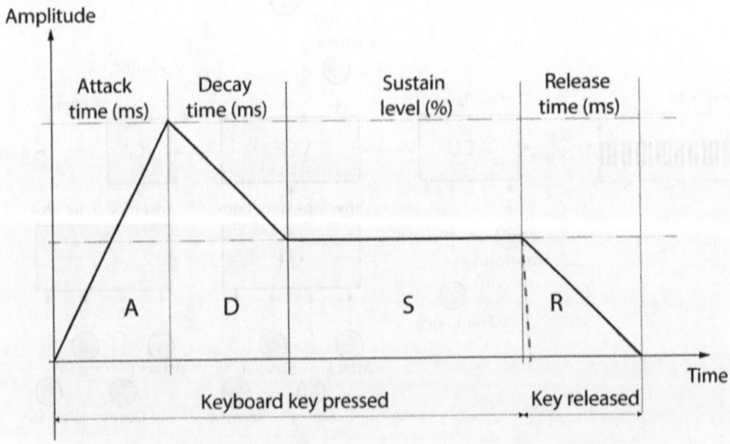

Figure 2.2. *The envelope used by the Minimoog Model D*

NOTE.– The Minimoog filter has three parameters: the cutoff frequency, the accentuation or emphasis and the amount of contour.

The configuration shown in Figure 2.3 simulates the sound of a French horn.

Instructions and comments are as follows:

– position the rotary knobs, switches and wheels as shown below;

– PITCH wheel: 0;

– MOD wheel: 0;

– TUNE: 0;

– GLIDE: OFF;

– DECAY: ON;

– OSCILLATOR MODULATION: OFF;

– OSCILLATOR 1: RANGE = 8' – WAVEFORM = sawtooth;

– OSCILLATOR 1: VOLUME: 5;

– MIXER switches: ON; OFF; OFF; OFF; OFF;

– FILTER MODULATION: OFF;

– KEYBOARD CONTROL 1: OFF;

– KEYBOARD CONTROL 2: OFF;

– FILTER: CUTOFF = 2; EMPHASIS = 0; CONTOUR = 4;

– envelope 1 (FILTER): ATTACK TIME = 370 ms; DECAY TIME = 300 ms; SUSTAIN LEVEL = 8%;

– envelope 2 (MODIFIERS): ATTACK TIME = 10 ms; DECAY TIME = 300 ms; SUSTAIN LEVEL = 9%;

– OUTPUT VOLUME: 5;

– MIX OUTPUT: ON.

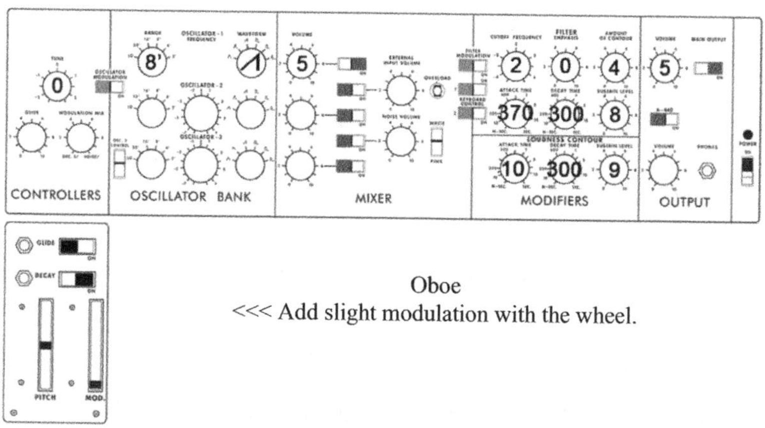

Figure 2.3. *Configuration of the Minimoog Model D to obtain the sound of a French horn (1 VCO, 1 VCF, 2 EG). For a color version of this figure, see www.iste.co.uk/reveillac/synthesizers2.zip*

2.1.2. *Behringer 2600 (ARP 2600)*

Like the Minimoog, this synthesizer has two envelope generators; the first, intended for the VCA by default, is based on the four classic ADSR parameters, and the second is based on the two AR parameters.

For this exercise, we will use the pre-routing of modules already established with the 2600.

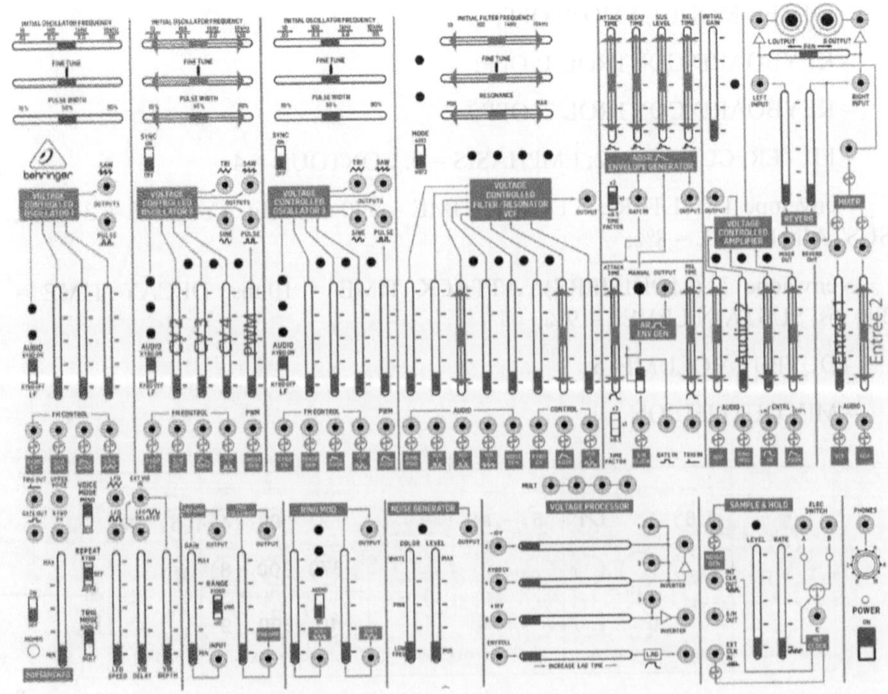

Figure 2.4. *Using the two envelope generators*

Instructions and comments are as follows:

– place the sliders and switches as shown on the patch in Figure 2.4;

– you can note the path of the signal, which starts from VCO2 to join the mixer before joining the VCF, controlled by envelope generator 2 (AR), ending with the VCA controlled by envelope generator 1 (ADSR);

– the envelope generators influence the ADSR slider of the VCF and the AR/ADSR of the VCA;

– the action of the VCF can also be managed within the VCA.

2.1.3. *Max/MSP*

This new exercise is designed around exercise 3 in Chapter 1 (section 1.3.2). However, several changes have been made to introduce you to new signal processing possibilities.

All slider objects have been replaced with dial objects followed by a scale object, making it possible to transform the range of the values of the input signal by applying an increase or decrease factor to them.

By default, a dial object has a range of values between 0 and 127; the scale objects placed at the output modify the ranges to cover adequate ranges, for example, from 10 to 10,000 for attack (A) and release (R).

Another modification already dealt with in section 1.4.1 is the replacement of midiin and xnotein objects with notein and / 127 objects. This modification makes it possible to detect the velocity and, therefore, when a key on the keyboard is pressed.

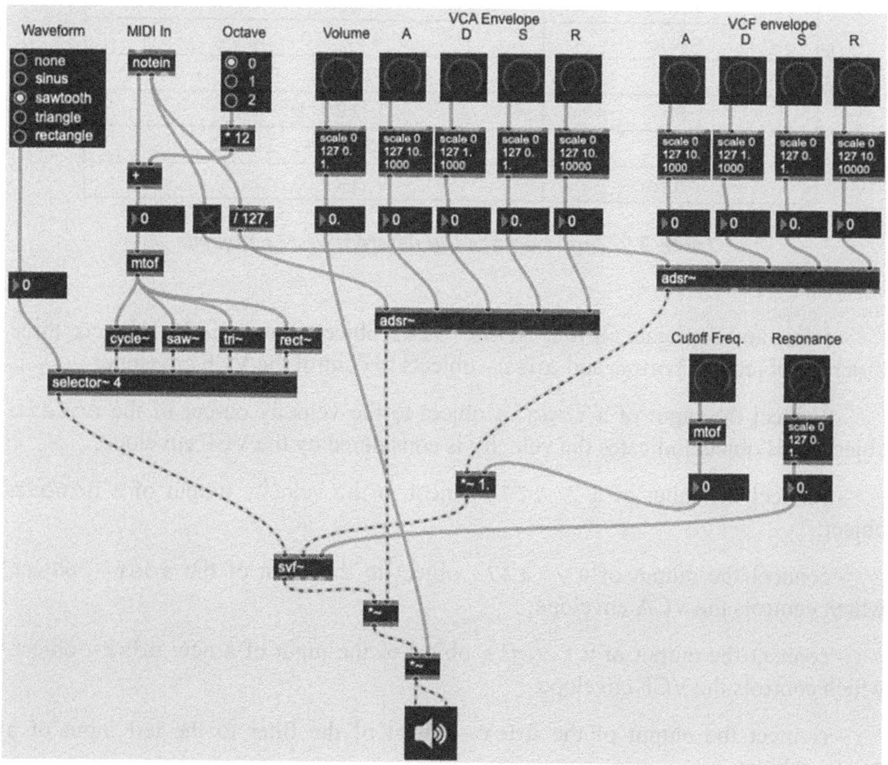

Figure 2.5. *Patch exercise 4 with Max/MSP*

The objects simulating the VCA envelope have been duplicated to be re-used as the VCF envelope.

Instructions and comments are as follows:

– repeat the patcher from exercise 3 (section 1.3.2);

– remove the 7 `slider` objects;

– replace all `slider` objects with `dial` objects with their default values, each followed by a `scale` object, except for the `Cutoff Frequency slider`, which will be followed by a `mtof` object, which will convert the values 0 to 127 from the `dial` object into a frequency (0–12,543 Hz);

– parameters of scale objects are given in Table 2.2;

Dial	**Scale arguments**
Volume	0 127 0. 1.
A	0 127 10. 1000
D	0 127 1. 1000
S	0 127 0. 1.
R	0 127 10. 10000
Resonance	0 127 0. 1.

Table 2.2. *Arguments of the different* `scale` *objects*

– select and duplicate all four ADSR `dial` objects, four `scale` objects, three `number` objects, `flonum` and `adsr~` objects to control the VCF envelope;

– connect the input of a `toggle` object to the velocity output of the `notein` object. This object indicates the velocity is considered by the VCF envelope;

– connect the input of a `/ 127.` object to the velocity output of a `notein` object;

– connect the output of a `/ 127.` object to the input of the `adsr~` object, which controls the VCA envelope;

– connect the output of a `toggle` object to the input of a new `adsr~` object, which controls the VCF envelope

– connect the output of the `adsr~` object of the filter to the left input of a `*~ 1.` object;

– connect the output of the `mtof` object dedicated to the cutoff frequency to a `number` object then to the right input of the `*~ 1.` object to multiply the shape of the envelope by the cutoff frequency;

– create an `svf~` filter object to replace the `lores~` filter object;

– connect the output of the object `*~ 1.` to the middle input, reserved for the cutoff frequency, of the filter object `svf~`;

– try the patcher and see the influence of each of the settings on the sound reproduction, particularly by acting on the settings of the filter envelope, its cutoff frequency and its resonance.

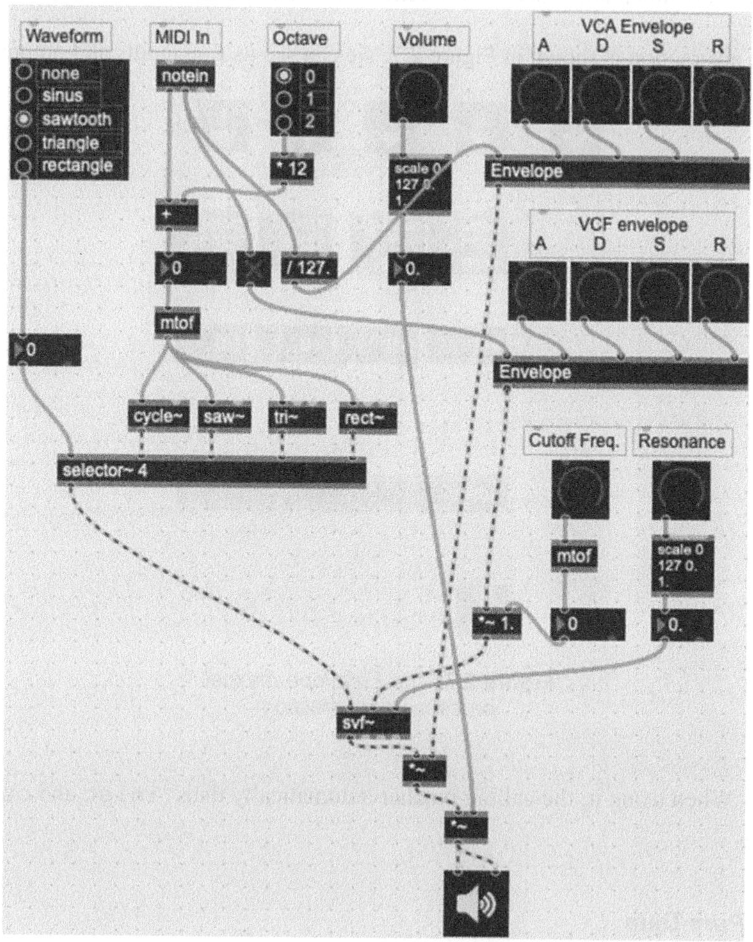

Figure 2.6. *Exercise 4 using the envelope abstraction object*

In Figure 2.6, a second patcher with an abstraction (external subroutine), called the Envelope, was created to instantiate an envelope twice and condense the development of the patcher. Its operation remains identical to the previous one.

NOTE.– Instead of an abstraction that requires two patchers, the program and its abstraction, it is possible to create a p (object subpatch – internal subroutine), under certain conditions, which will be integrated twice in the same patcher (consult the Max/MSP documentation, see the reference section at the end of the book).

The abstraction is shown in Figure 2.7; we can see its five inputs and output.

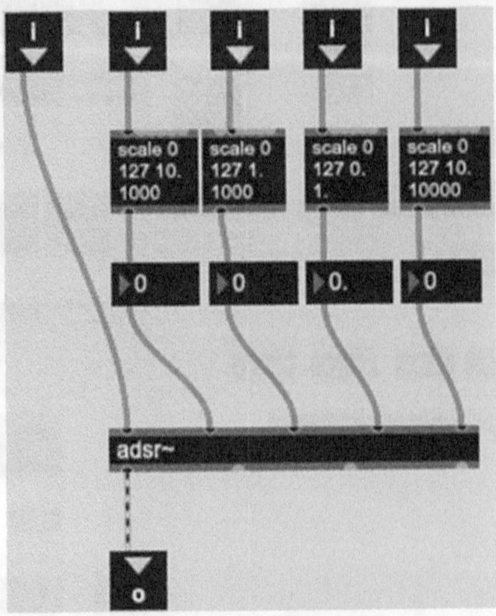

Figure 2.7. *The Envelope.maxpat processing abstraction*

NOTE.– When using it, the calling patcher automatically dials `inlet` and `outlet` objects.

2.1.4. *Pure Data*

We resume exercise 3 from Chapter 1 (section 1.3.3), to add this new envelope.

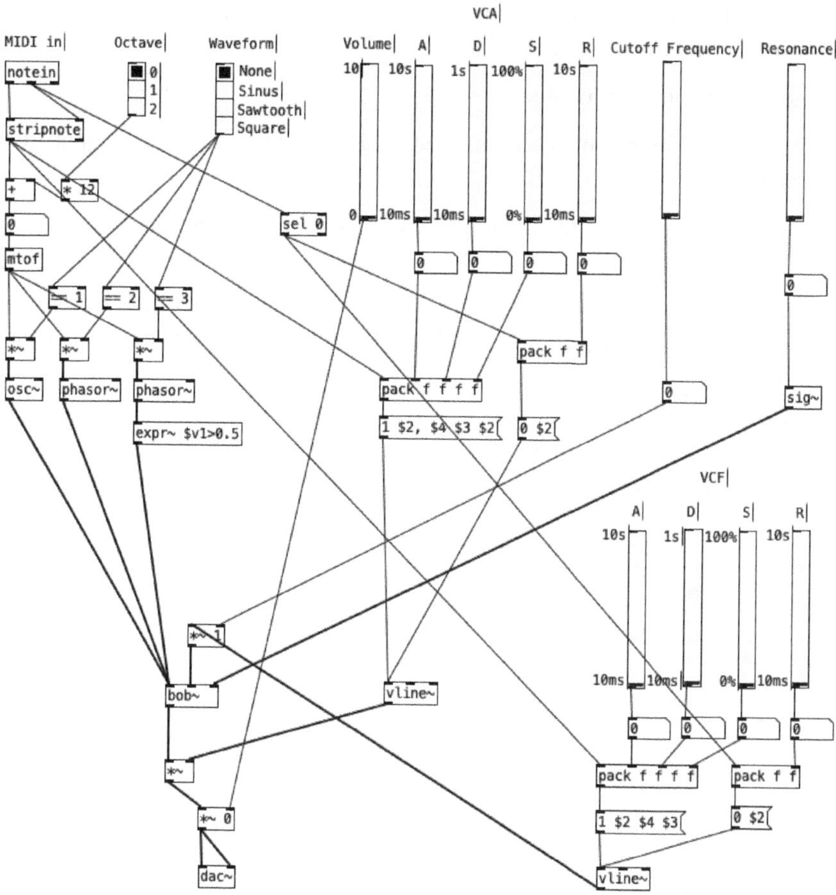

Figure 2.8. *Exercise 4 with its two envelopes, one for the VCA and one for the filter with Pure Data*

Instructions and comments are as follows:

– duplicate the objects involved in the VCA envelope (four `slider` objects, four `number` objects, two `pack` objects, two `message` objects and one `vline~` object);

– connect the left output of the `sel 0` object to the input of the duplicated `pack f f f f` object;

– connect the left output of the `stripnote` object to the input of the duplicated `pack f f` object;

– delete the `sig~` object connected to the `number` object that follows the `slider Cutoff Frequency` object;

– add a signal multiplier object `*~ 1`;

– connect the output of the duplicated `vline~` object to the left input of the `*~ 1` object;

– connect the output of the `number` object placed behind the `slider Cutoff Frequency` object at the input of the `*~ 1` object;

– connect the output of the `*~ 1` object to the middle input of the `bob~` object to consider the cutoff frequency and the ADSR parameters of the envelope intended for the filter;

– experiment with different VCA and VCF envelope settings according to the cutoff frequency and resonance of the filter. Be careful, the settings are sensitive, and you must find the right balance of filter parameters for both envelopes.

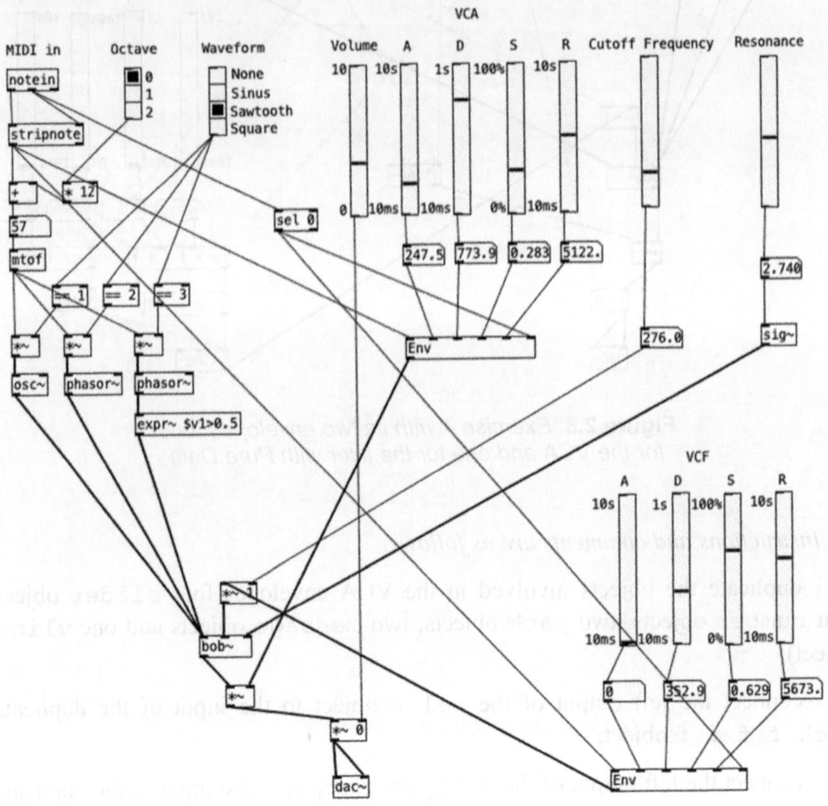

Figure 2.9. *Exercise 4 with the Env. abstraction object*

Figure 2.9 shows the same exercise replacing the envelope generator objects (two `pack` objects, two `message` objects and one `vline~` object) with an abstraction (external subroutine) in a file named Env.pd.

Figure 2.10 shows the `Env.pd` abstraction.

NOTE.– The `number` objects placed in the abstraction are redundant and not necessary. They ensure that the parameters of the `slider` objects are used correctly. A tilde (~) must follow an `outlet` object because it transmits a signal, not a value.

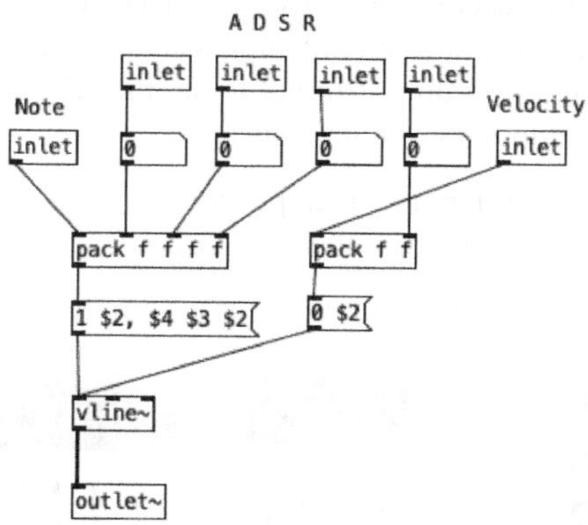

Figure 2.10. *The abstraction Env.pd for exercise 4*

2.1.5. *VCV Rack*

Continuing with our virtual modular synthesizer, we will resume exercise 3 from Chapter 1 (section 1.3.4) and add an additional ADSR module.

Instructions and comments are as follows:

– resume exercise 3;

– add an ADSR module (fundamental blocks library) between the VCO and VCF modules;

– connect the GATE output of the MIDI-CV module to the GATE input of the new ADSR module;

– connect the ENV output of the ADSR module to the CUT input of the VCF module;

– move the cable from the high-pass filter (HPF) output of the VCF to the low-pass filter (LPF) output to give a more apparent effect of the action of the envelope on the filter cutoff frequency. The HPF output may also be suitable in some cases depending on the desired effect;

– for the envelope generator to be sensitive on the VCF, turn the CUTOFF button to its minimum;

– adjust the action level at the VCF input by turning the CUT button close to its maximum. Later, after a few tries, you can reduce its intensity, just like increasing the CUTOFF level;

– the ADSR envelope generator can also be connected to the RES (resonance) or DRIVE input at the VCF output. Experiment to see the influence of the envelope on these parameters. In this case, do not forget to set the RES and DRIVE buttons to low values and the input dosages to high values;

– remember to draw different envelope curves for the VCA and the VCF; otherwise, the actions of one and the other will be confused.

Figure 2.11. *Exercise 4 for the VCV Rack. For a color version of this figure, see www.iste.co.uk/reveillac/synthesizers2.zip*

2.2. Exercise 5 – integrating an LFO

An LFO is present in most synthesizers. Some may even have several (2 on Korg M1, 2 on an Arturia MatrixBrute, 6 per patch on a Roland D-50, etc.). An LFO can be compared to a periodic envelope.

It generates the modulations essential for creating specific signals and is generally applied to VCAs to create a tremolo effect and to VCOs to generate a vibrato effect. The LFO can also be the basis of more sophisticated effects such as the phaser or the chorus, but its functions do not stop there; it can perform pulse wave modulation (PWM), modulate the filter cutoff frequency, the resonance, the amplitude, the pan, the frequency of another LFO, etc.

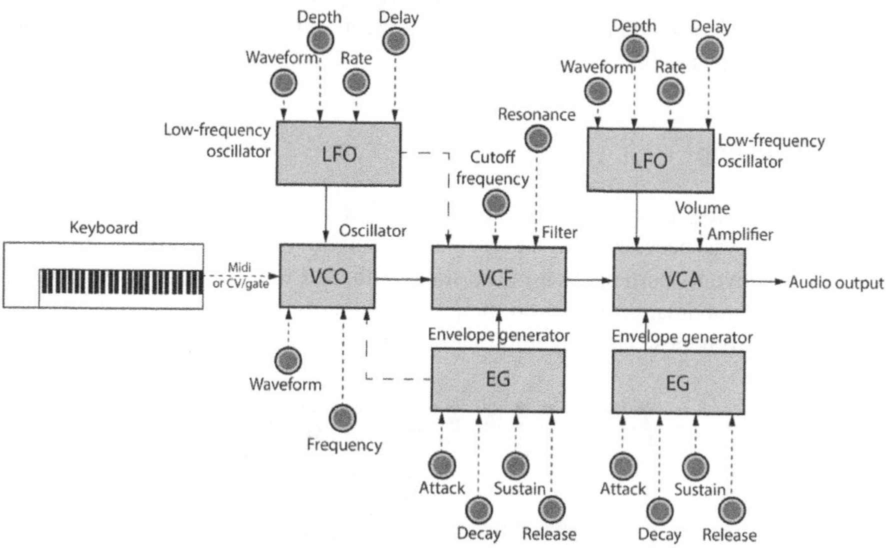

Figure 2.12. *Subtractive sound synthesis with one VCO, two EGs, two LFOs and a VCA. The command signals are in dotted lines*

It has numerous parameters; the first four are encountered most often:

– the waveform (sine, sawtooth, square, triangle, random, etc.);

– the rate to obtain a rapid modulation, generally from 0.1 to 20 Hz, but it can be much higher up to 250 Hz (Bit 01/99, for example);

– the depth to control the action of the LFO on the primary signal. It is sometimes called the direct modulation rate;

– the delay time to delay its appearance when pressing a key on the keyboard or when triggering it. On some synthesizers, it can be influenced by the pitch of the note or the rate (initial pressing of the note, speed at which a key or pad is pressed);

– MIDI synchronization is very useful for creating rhythm;

– humanization, which adds small random variations;

– the polarity, which inverts the LFO phase. This polarity can, on certain synthesizers, be adjusted according to a determined angle (Yamaha SY77, for example);

– the envelope, which can control the evolution of the LFO modulation rate over time;

– indirect modulation rate, which defines the depth according to the action of a controller such as a modulation wheel or even the aftertouch (pressing a key or pad while sustaining a note).

2.2.1. *Minimoog*

The Minimoog is a particular case; the original Model D omitted an LFO. Therefore, it was necessary to use oscillator 3 in Lo mode, deactivate keyboard tracking, remove its participation in the audio path, and route it, via the Controllers section, to the modulation wheel, to the pitch oscillators 1 and 2, at the filter cutoff frequency or even these last two at the same time.

a) b)

Figure 2.13. *Minimoog: (a) classic Model D and (b) reissued Model D. For a color version of this figure, see www.iste.co.uk/reveillac/synthesizers2.zip*

Figures 2.13 and 2.14 show two examples (oboe and heartbeat) using the Lo mode on the original Minimoog Model D (1971 to 1981), followed by Figure 2.15 with an example of a more recently reissued Minimoog Model D (2016), to which an LFO was added. Its control (LFO Rate) is above the pitch and modulation wheels (Figure 2.13).

The control switch can be lifted to select the LFO waveform, which is triangular or square, by default.

NOTE.– These sounds come from the different manuals that come with the Minimoog when purchased.

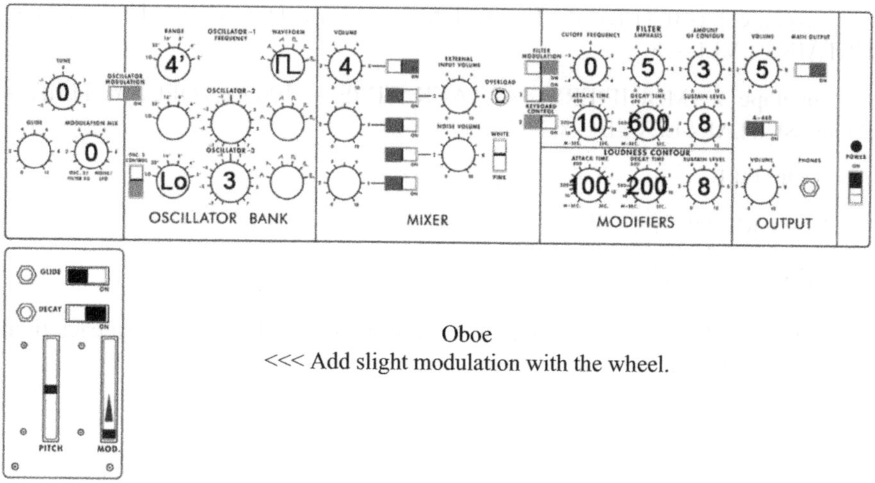

Figure 2.14. *Configuring the Minimoog Model D to obtain an oboe sound. For a color version of this figure, see www.iste.co.uk/reveillac/synthesizers2.zip*

Instructions and comments are as follows:

– position the switches and knobs as shown in Figure 2.14;

– PITCH knob: 0;

– MOD knob: > 0;

– GLIDE: OFF;

– DECAY: ON;

– TUNE: 0;

– MODULATION MIX: 0;

– OSCILLATOR MODULATION: ON;

– OSCILLATOR 1: 4’ – WAVEFORM = square;

– OSC.3 CONTROL: ON;

– OSCILLATOR 3: RANGE = LO;

– OSCILLATOR 1 VOLUME = 4;

– MIXER switches: ON; OFF; OFF; OFF; OFF;

– FILTER: CUTOFF = 0; EMPHASIS = 5; CONTOUR = 3;

– envelope 1 (FILTER): ATTACK TIME = 10 ms; DECAY TIME = 600 ms; SUSTAIN LEVEL = 8;

– envelope 2 (MODIFIERS): ATTACK TIME = 100 ms; DECAY TIME = 200 ms; SUSTAIN LEVEL = 8;

– A440: OFF;

– VOLUME OUTPUT: 5;

– MIX OUTPUT: ON.

Figure 2.15 shows the configuration to imitate the sound of a heartbeat each time a key is pressed on the keyboard.

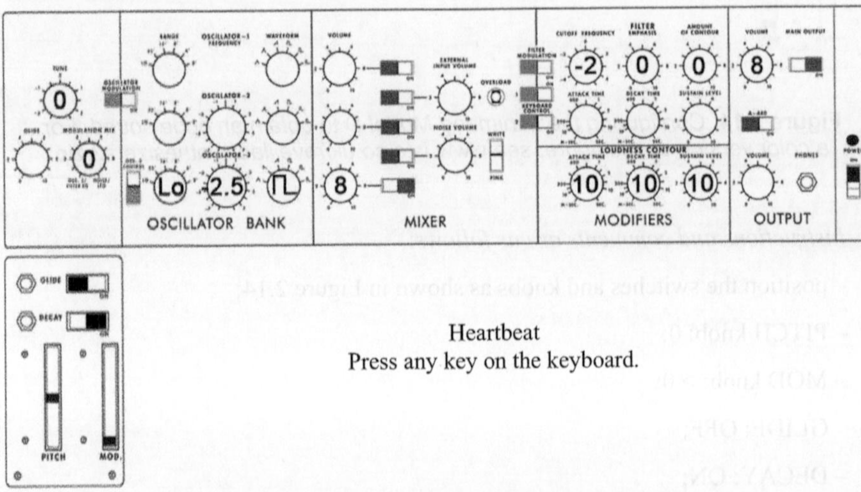

Figure 2.15. *Configuring the Minimoog Model D to obtain a heartbeat sound. For a color version of this figure, see www.iste.co.uk/reveillac/synthesizers2.zip*

Instructions and comments are as follows:

– position the switches and knobs as shown in Figure 2.15;

– PITCH knob: 0;

– MOD knob: 0;

– GLIDE: OFF;

– DECAY: ON;

– TUNE: 0;

– MODULATION MIX: 0;

– OSCILLATOR MODULATION: OFF;

– OSC.3 CONTROL: ON;

– OSCILLATOR 3: RANGE = LO; –2.5; WAVEFORM = square;

– OSCILLATOR 3: VOLUME = 8;

– MIXER switches: OFF; OFF; OFF; OFF; ON;

– FILTER: CUTOFF = –2; EMPHASIS = 0; CONTOUR = 0;

– envelope 2 (MODIFIERS): ATTACK TIME = 10 ms; DECAY TIME = 10 ms; SUSTAIN LEVEL = 10;

– A440: OFF;

– VOLUME OUTPUT: 8;

– MIX OUTPUT: ON.

Figure 2.16 presents a configuration adapted to the reissued Minimoog Model D of 2016 to obtain a sound close to the voice of a robot. It uses the new LFO controls available for this version of the Minimoog.

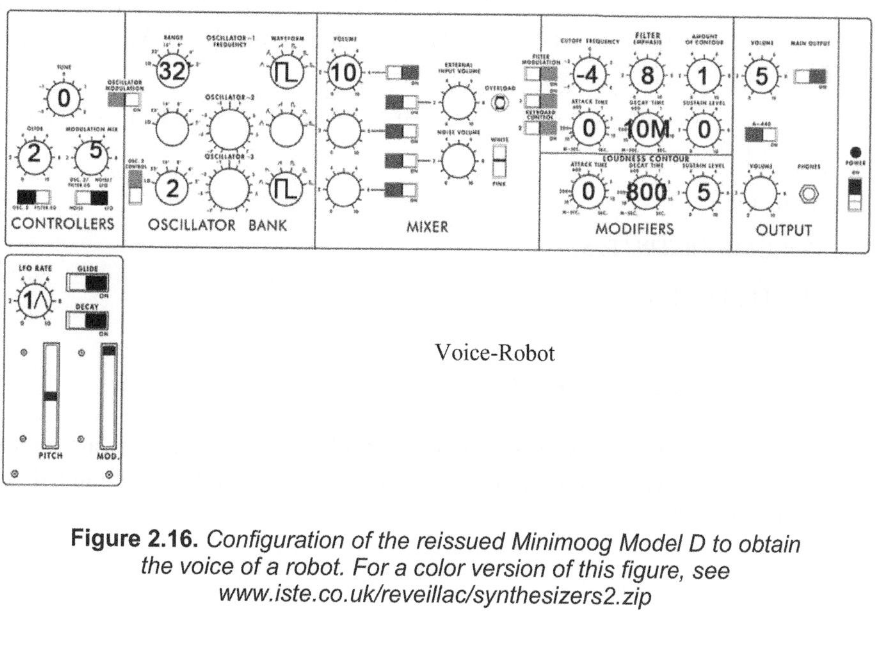

Figure 2.16. *Configuration of the reissued Minimoog Model D to obtain the voice of a robot. For a color version of this figure, see www.iste.co.uk/reveillac/synthesizers2.zip*

Instructions and comments are as follows:

– position the switches and knobs as shown in Figure 2.16;

– PITCH knob: 0;

– MOD knob: 0;

– GLIDE: ON;

– DECAY: ON;

– TUNE: 0;

– GLIDE: 2;

– MODULATION MIX: 5;

– OSCILLATOR MODULATION: OFF;

– OSC.3 Control: OFF;

– OSCILLATOR 1: 32' – WAVEFORM = square;

– OSCILLATOR 1: VOLUME = 10;

– OSCILLATOR 3: 2' – WAVEFORM = square;

– MIXER switches: ON; OFF; OFF; OFF; OFF;

– FILTER: CUTOFF = – 4; EMPHASIS = 8; CONTOUR = 1;

– envelope 1 (FILTER): ATTACK TIME = 0 ms; DECAY TIME = 10 ms; SUSTAIN LEVEL = 0;

– envelope 2 (modifiers): ATTACK TIME = 0 ms; DECAY TIME = 800 ms; SUSTAIN LEVEL = 5;

– A440: OFF;

– VOLUME OUTPUT: 5;

– MIX OUTPUT: ON.

2.2.2. *Novation Bass Station II*

This small synthesizer has many features that allow you to explore a vast sound palette; unfortunately, its weak point remains its controls, which are limited to stringent requirements.

Many settings pass through selectors, which reduce the number of controls on the front panel. For example, the seven oscillator adjustment parameters exist only once (five knobs and two push buttons) and a selector allows you to choose between VCO1 and VCO2.

Other settings are accessible via a function button, pressing a key on the keyboard and two entry buttons for the chosen value; effective, but not very ergonomic.

The display of values assigned to most of the settings is through a single, small alphanumeric screen with light-emitting diodes, showing three characters, which is not easy to read.

It is therefore impossible to materialize a configuration (patch) in the form of a diagram of the front panel.

As a workaround, we have chosen to use the open source online editor "Bass Station 2 Web Interface 2.11.0" from Studiocode.dev, under GNU license, developed by the Swiss engineer François Georgy[1].

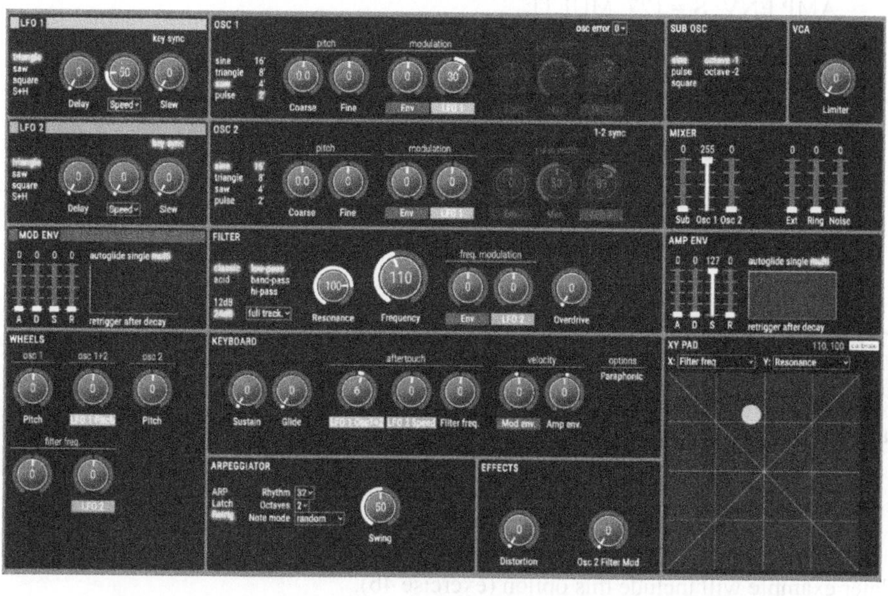

Figure 2.17. *Police siren on Novation Bass Station II. For a color version of this figure, see www.iste.co.uk/reveillac/synthesizers2.zip*

1 Available at: www.studiocode.dev/bs2-editor/.

It offers an ergonomic and easy-to-use online user interface, working with the Google Chrome internet browser.

The Novation Bass Station II configurations will all be presented via screenshots of this software.

For this first exercise with the Novation Bass Station II, here is a configuration generating the sound of an American police siren. It involves a sawtooth oscillator, an LFO, filtering and an appropriate envelope for the VCA.

Instructions and comments are as follows:

– adjust the various controls as indicated below;

– LFO 1: SPEED = 50;

– OSC 1: TRIANGLE 2';

– OSC 1: MODULATION LFO 1 = 30;

– FILTER: RESONANCE = 100, FREQUENCY = 110;

– AMP ENV: S = 127, MULTI;

– MIXER: OSC 1 = 255;

– SUB OSC: by default;

– ARPEGGIATOR: by default;

– all other parameters: 0;

– remember to set the output volume to the correct value.

2.2.3. *Max/MSP*

For this exercise, the Max/MSP patcher includes two LFOs, one for the vibrato and a second for the tremolo.

Users can choose the effect – vibrato or tremolo – and adjust their rate and depth.

Initially, this patcher (exercise 4a) is not intended for use with a keyboard. A later example will include this option (exercise 4b).

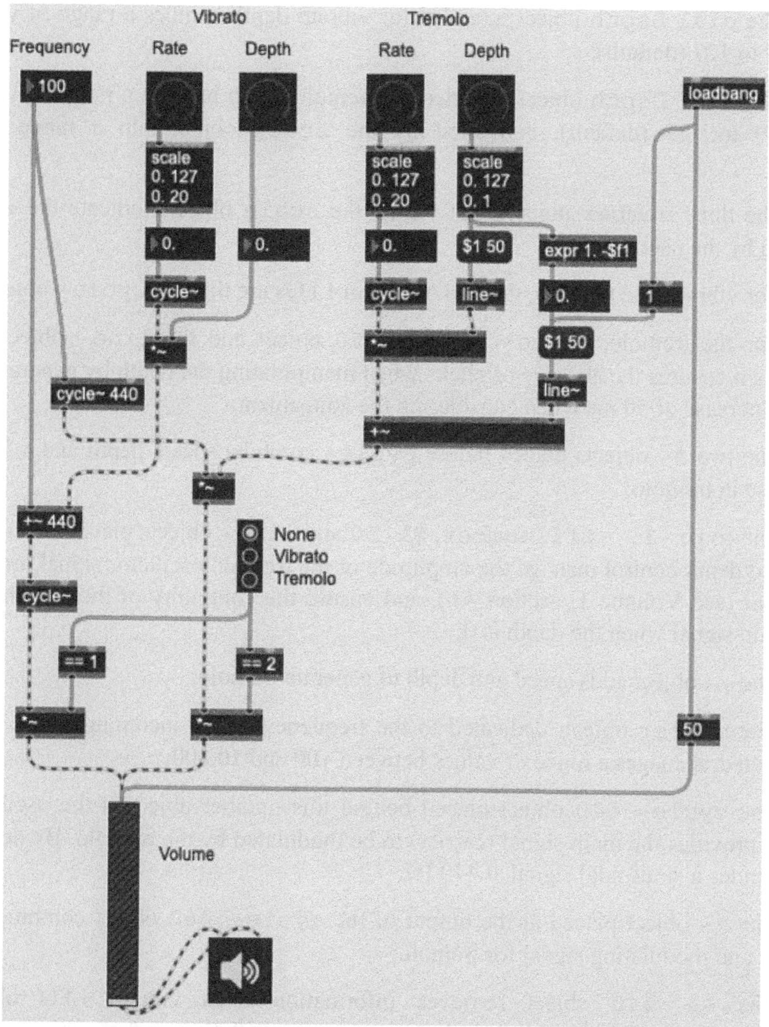

Figure 2.18. *Exercise 4a: vibrato and tremolo with Max/MSP*

Instructions and comments are as follows:

– the four `dial` objects, three of which are followed by `scale` objects, are the vibrato and tremolo controls;

– the two `dial` and two `scale` objects for the vibrato and tremolo rates manage a range of values from 0 to 127 (by default), transformed by the `scale` into a range of 0–20;

– the dial Depth object intended for vibrato depth handles a range of values from 0 to 127 (default);

– the dial Depth object intended for tremolo depth handles a range of values from 0 to 127 (default), converted by the scale object into a range from 0 to 1;

– the three number objects that follow the scale objects indicate the values chosen by the user;

– for vibrato and tremolo, the two oscillator LFOs are the two cycle~ objects;

– for the tremolo, the message $1 50 object and the line~ object that follows it ensures the absence of clicks when manipulating the depth by generating a small increase of 50 ms when considering the adjustment;

– the two *~ objects placed before cycle~ combine speed, depth and a 50 ms increase in tremolo;

– the expr~1. -$f1, number, $1 50 and line~ objects placed behind the tremolo depth control manage the amplitude of the tremolo oscillator signal, make it unipolar (see Volume 1, section 3.6), and ensure the continuity of the presence of the main signal when the depth is 0;

– the +~ object adds speed and depth to generate tremolo;

– the number object, dedicated to the frequency of the incoming signal to be modulated, manages a range of values between 100 and 10,000;

– the cycle~ 440 object placed behind this number object is the oscillator, which provides the main signal (carrier) to be modulated by the tremolo. By default, it generates a sinusoidal signal at 440 Hz;

– the *~ object placed at the output of the cycle~ 440 object combines the carrier and modulating signal for tremolo;

– the +~ 440 object retrieves information from vibrato LFO objects (modulating signal) to send it from its output to an oscillator cycle~ object, which provides the main (carrier) signal dedicated to vibrato. By default, this cycle~ object will generate a sinusoidal signal at 440 Hz;

– the five == 1, == 2, radiogroup, None, Vibrato, Tremolo and *~ (x 2) objects ensure the effect the user chooses;

– a gain object manages the overall volume;

– a loadbang object preloads, via two int objects, the value 1 to the $1 50 tremolo message object and 50 to the gain object to ensure the presence of a signal at start-up and the correct volume as soon as the user chooses the vibrato or tremolo effect;

– manipulate the two LFOs and compare the effects on the main signal. Vibrato is a constant modulation of the pitch (frequency) of the sound that oscillates around its midpoint. Tremolo is a volume modulation.

Let us now modify exercise 4a so that our patcher recognizes a keyboard on a MIDI in input.

You will find in Figure 2.19 the modifications to be made.

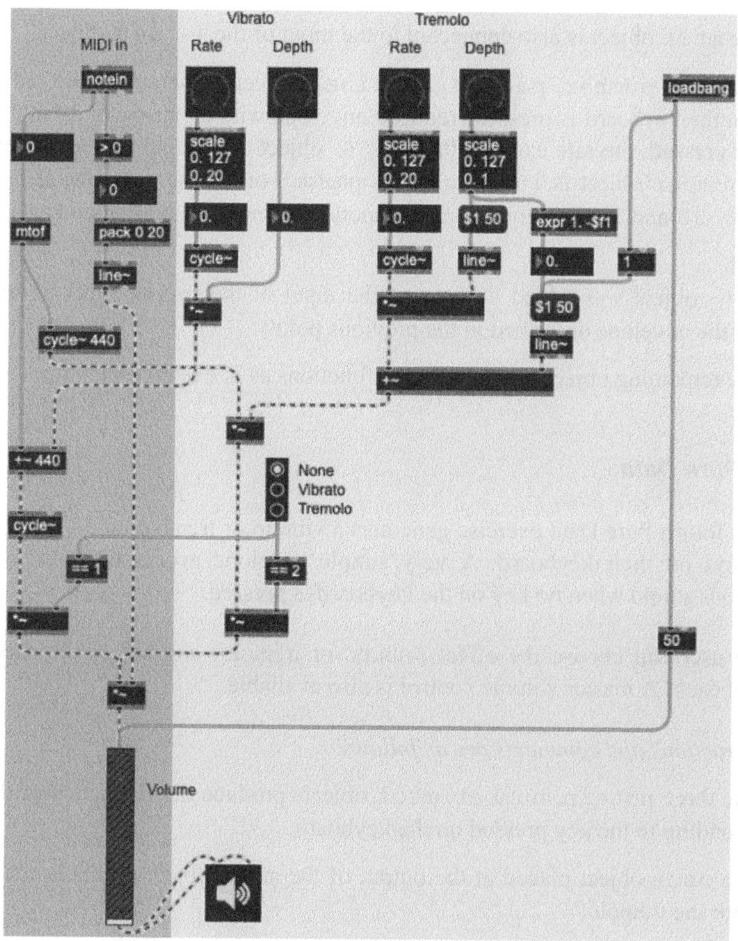

Figure 2.19. *Exercise 4b: LFO and MIDI keyboard*

Instructions and comments are as follows:

– only the grayed part on the left of Figure 2.19 has been modified compared to exercise 4a;

– a `notein` object was added to handle a MIDI keyboard;

– the `number` object that follows it is there to check the number of the transmitted MIDI note;

– the `mtof` object transforms the note into a frequency, and its output is connected to the input of the `cycle~ 440` object, one of the oscillators generating the primary signal to be modulated;

– the `mtof` object is also connected to the input of the `+~ 440` object;

– the `> 0`, `number`, `pack 0 20`, `line~` objects generate a note only when a key on the keyboard is pressed (reduced envelope with an attack of 20 ms). When a key is pressed, the rate exceeds 0; the `> 0` object carries out this test. The value of the `number` object is 1 when a key is pressed; otherwise, its value is 0 for the `pack 0 20` and `line~` objects; they generate an increase in attack when a note is pressed;

– a `*~` object was added in front of the input of the `gain` object to take into account the envelope described in the previous point;

– the remaining objects have the same functions as in exercise 4a.

2.2.4. *Pure Data*

This fourth Pure Data exercise generates a vibrato or tremolo on all the notes the user plays on their keyboard. A very simple envelope avoids the presence of a continuous sound when no key on the keyboard is pressed.

The user can choose the effect (vibrato or tremolo) and adjust the speed and depth of each. A master volume control is also available.

Instructions and comments are as follows:

– the three `notein`, `number`, `mtof` objects produce the frequency of the note corresponding to the key pressed on the keyboard;

– the `osc~` object placed at the output of the `mtof` object produces the carrier signal for the tremolo;

– the four > 0, number, pack 0 20 and line~ objects, connected behind the rate output to the right of the notein object, generate a minimalist envelope with a 20 ms attack when pressing a key, detected by a rate greater than 0;

– the five vradio (none, tremolo, vibrato) objects, == 1, == 2 followed by two multiplier objects *~ allow you to choose the type of effect, vibrato or tremolo, depending on the user selection (0: none, 1: tremolo, 2: vibrato);

– the slider object intended for the vibrato depth controls a range of linear values from 0 to 127;

– the slider object intended for vibrato speed has a linear range of values from 0 to 20;

– the slider object intended for tremolo depth has a linear range of values from 0 to 0.5;

– the slider object intended for tremolo speed has a linear range of values from 0 to 20;

– the two number objects that follow the slider objects indicate the values chosen by the user;

– for the vibrato and tremolo, the oscillators in the two LFOs are the two osc~ objects;

– the two *~ objects placed behind osc~ combine speed and an increase in 50 ms for tremolo;

– for the tremolo, the pack f 50 object and line~ object control the amplitude, which defines the depth of the tremolo, depending on the value sent by the slider Depth object;

– the expr~ 1 - $v1 object placed behind the line~ object makes the signal unipolar (otherwise, we would have ring modulation instead of tremolo, which is an amplitude modulation (see Volume 1, section 3.6);

– the +~ adds speed and depth to generate tremolo;

– for vibrato, the +~ adds the value of the mtof output frequency and the vibrato modulating signal to direct it towards the osc~, which produces the carrier signal for the vibrato;

– the two *~ objects placed behind the *~ objects, intended for effect chosen by the user, combine the carrier signal and modulating signal for vibrato and tremolo;

– the four objects slider volume, / 127, *~ 0.5 and dac~ ensure the output volume of the audio signal is transmitted and controlled.

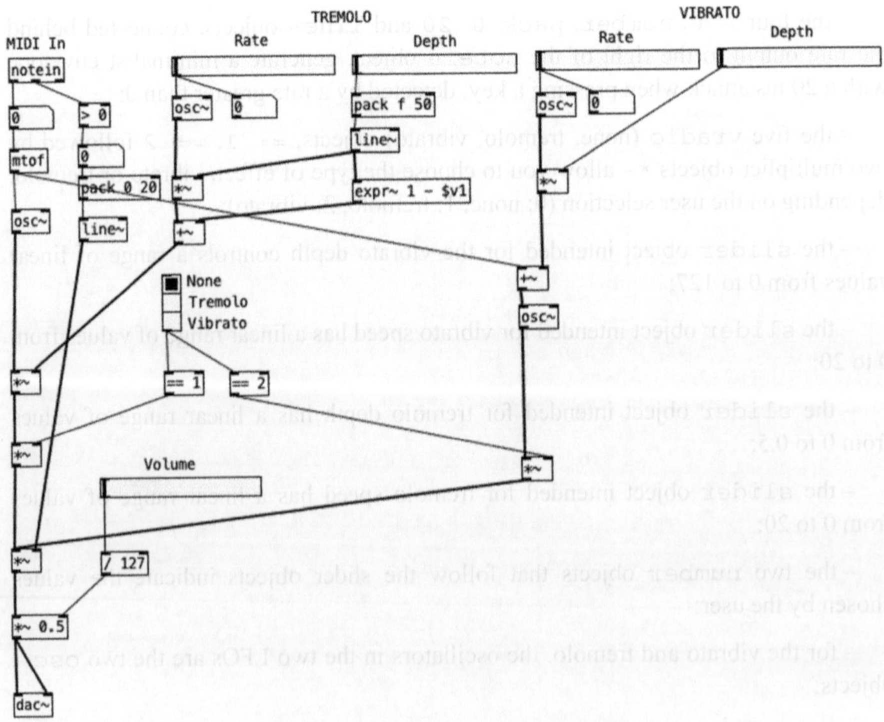

Figure 2.20. *Exercise 4: two LFOs for a vibrato and a tremolo in Pure Data*

2.2.5. *VCV Rack*

To complete the work already carried out in the exercises covered, we will add an LFO in the modular synthesizer we built during the previous exercises.

A VCA MIX module will also be added to correctly mix the signals and ensure a balance (wet/dry) between the modulating signal and the modulated signal (carrier).

Instructions and comments are as follows:

– return to the module built in exercise 4 (see section 2.1.5);

– add a VCA MIX module and an LFO module (fundamental blocks library);

– move the cable connecting the output of the ADSR module, dedicated to the envelope, to the IN 1 input of the VCA MIX module;

– connect the SIN output of the LFO module to the IN 2 input of the VCA MIX module;

– connect the MIX output of the VCA MIX module to the VCA module control input;

– you can adjust the influence of the LFO and the volume of the carrier signal with the sliders present on inputs 1 and 2 of the VCA MIX module;

– the modulation speed can be adjusted using the FREQ button on the LFO module;

– you can also move the cable at the LFO module output to obtain other signal shapes (TRI – triangular, SAW – sawtooth, SQR – square).

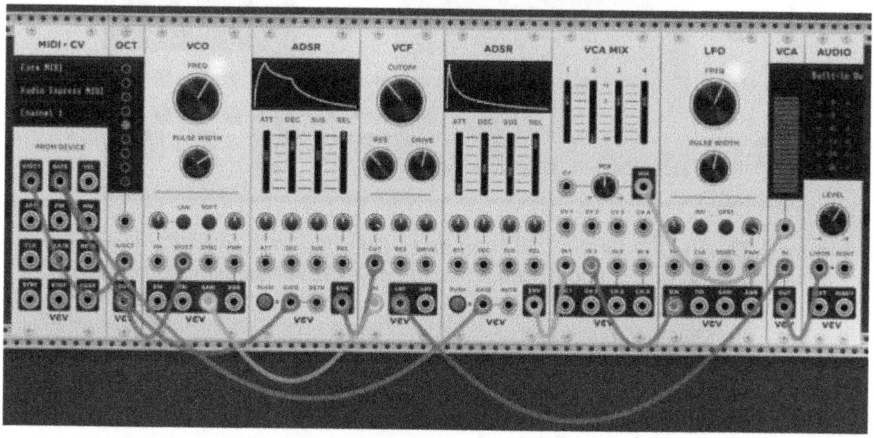

Figure 2.21. *An LFO for VCA in a VCV Rack. For a color version of this figure, see www.iste.co.uk/reveillac/synthesizers2.zip*

To continue, we will place a second LFO, which will this time act on the VCO. An oscilloscope will also be implemented to visualize the different signals available at the output. It is a practical and effective tool to have a real vision of the waveforms generated concerning the sound rendering obtained or to be obtained.

It is often difficult to understand what is happening within a module; moving the oscilloscope to different control points lets you see a signal's evolution or presence. Furthermore, this oscilloscope has two inputs, an excellent way to compare signals.

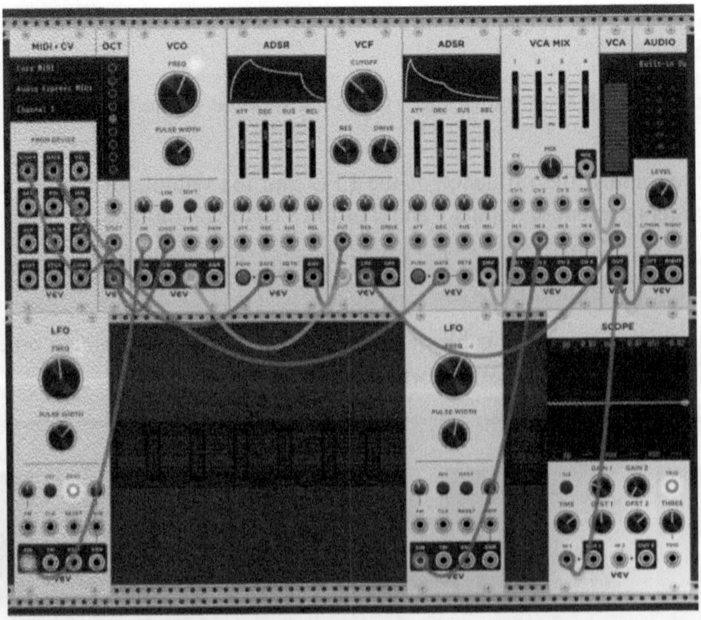

Figure 2.22. *An LFO for the VCO and an LFO for the VCA in a VCV Rack. For a color version of this figure, see www.iste.co.uk/reveillac/synthesizers2.zip*

Instructions and comments are as follows:

– for greater clarity, the two LFO modules and the oscilloscope module have been placed on a second row of the rack;

– add a second LFO module;

– add a SCOPE module (fundamental blocks library);

– connect the IN 1 input of the SCOPE module to the OUT output of the VCA module;

– connect one of the outputs, SIN, TRI, SAW or SQR of the new LFO module to the FM input of the VCO module;

– by acting on the second linear volume potentiometer (faders) of the VCA MIXER module, you can remove the modulation of the second LFO to leave only that of the first, which directly controls the VCO;

– you will notice that the first LFO acts like a tremolo, it cyclically modifies the sound pitch (frequency modulation – FM), while the second generates a vibrato effect, a periodic modulation of the volume. Very different waveforms are visible on the SCOPE module screen when it is set correctly (TIME, the time base, and GAIN, the amplitude of the signal);

– mixing the signals from the two LFOs applied to the VCO can give surprising effects; do not hesitate to try very different combinations of parameters on all modules. Listen to the sound reproduction while analyzing the different waveforms present on the oscilloscope control screen;

– you can also move the cables connected to the SCOPE module to other points or add a second one to compare two signals. The offset settings (OFST1 and OFST2) will allow you to superimpose both curves.

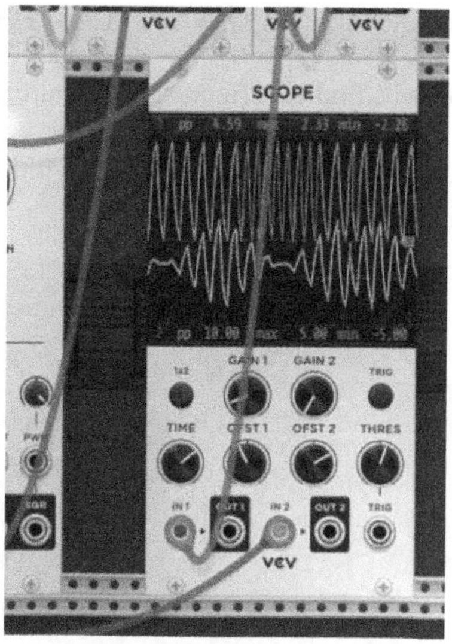

Figure 2.23. *The SCOPE module control screen has two signals, the VCO output (top) with a frequency-modulated signal and the VCA output (bottom) with an amplitude-modulated signal. For a color version of this figure, see www.iste.co.uk/reveillac/synthesizers2.zip*

2.2.6. *Reaktor*

Reaktor is a virtual modular synthesizer like the VCV Rack, although some of its functionalities are more elaborate. It is important to note that as with hardware synthesizers, although functionally similar, it is not possible to copy and paste between the two. The sound reproduction is different, the modules are distinct, and the connections established between the different modules often differ. The input–output points are also dissimilar, with the VCV Rack being closer to reality and the

Eurorack standard. At the same time, Reaktor is more focused on a proprietary system belonging to Native Instruments.

The fact remains that Reaktor is an excellent modular synthesizer offering many possibilities, which you will discover via the following exercises.

The exercises were all done using version 6 of the Native Instruments software. A less recent version may still be suitable for adequately executing the work, even if some adaptations will be necessary.

In this first exercise dedicated to this virtual synthesizer, we will set up a configuration like other synthesizers, VCO, VCF, VCA, LFO and GE.

For each exercise, you will find two figures, a structure mode and a panel mode view of all the modules. Depending on your preferences, you can decide to work in one mode or the other or even both. There is even, within Reaktor, a split mode allowing you to view both modes side by side.

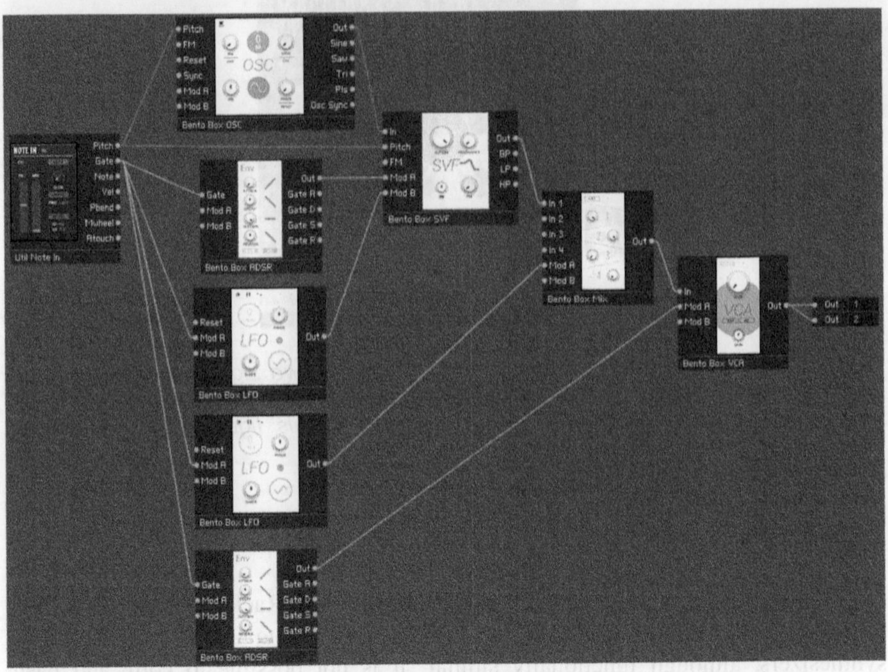

Figure 2.24. *The different Reaktor blocks in structure mode. For a color version of this figure, see www.iste.co.uk/reveillac/synthesizers2.zip*

Instructions and comments are as follows:

– create a new rack (.nksr);

– keep the modules IN, NOTE IN and OUT;

– choose the BENTO BOX subfolder and add the following blocks: OSC-Bento Oscillator, FLT-Bento SVF Filter, AMP-Bento VCA, MIX-Bento Mix, 2 × MOD-Bento LFO and 2 × MOD-Bento ADSR Envelope;

– then connect the different blocks together as shown in Figure 2.24 in accordance with Table 2.3. Comments explain the role of each connection;

Block	Out	Block	In	Comments
NOTE IN	Pitch	OSC	Pitch	Sends the MIDI note from the keyboard to the oscillator
NOTE IN	Pitch	SVF	Pitch	Provides keyboard tracking for the filter
NOTE IN	Gate	ENV 1	Gate	Triggers envelope 1
NOTE IN	Gate	LFO 1	Mod A	Triggers LFO 1
NOTE IN	Gate	LFO 2	Mod A	Triggers LFO 2
NOTE IN	Gate	ENV 2	Gate	Triggers envelope 2
OSC	Out	SVF	In	Sends the signal to the filter block input
ENV 1	Out	SVF	Mod A	Allows envelope 1 to control the SVF filter on input A
LFO 1	Out	SVF	Mod B	Allows LFO 1 to control the filter on its input B
LFO 2	Out	MIX	Mod A	Allows LFO 2 to modulate the signal on the MIX block from input A
ENV 2	Out	VCA	Mod A	Allows envelope 2 to control the VCA on input A
SVF	Out	MIX	In 1	Sends the filtered signal to the MIX block
MIX	Out	VCA	Mod A	Sends the signal to the VCA block
VCA	Out	OUT	Out 1	Sends the signal to audio output 1
VCA	Out	OUT	Out 2	Sends the signal to audio output 2

Table 2.3. *Connections to be made on the Reaktor rack*

– configure modulation buses A and B using appropriate sliders;

– for the LEVEL knob of the VCA, and the CUTOFF and RESONANCE knobs of the SVF filter, the slider of modulation bus A must be moved upwards (see Figure 2.25);

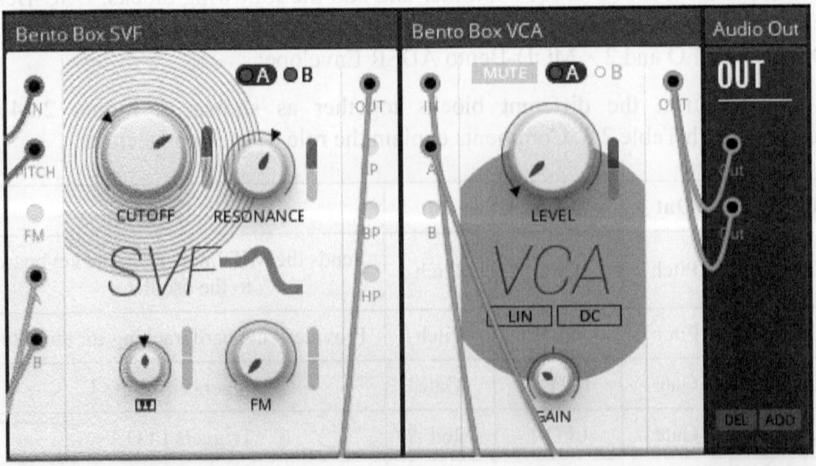

Figure 2.25. *Detail of the SVF and VCA modules on the panel. The different sliders of modulation bus A can be seen. For a color version of this figure, see www.iste.co.uk/reveillac/synthesizers2.zip*

– for the CUTOFF and RESONANCE buttons of the SVF filter, the slider of modulation bus B must be moved down;

– for button 1 of the MIX block, the slider of modulation bus A must be moved up;

– the MIX block mixes the modulation of LFO 2 with the carrier signal;

– LFO 1 modulates the SVF filter envelope;

– you can also act on modulation bus A of each LFO by moving their slider upwards. The GATE signal from the NOTE IN block will be considered in this case.

Try each block's possible settings: waveform, cutoff frequency, resonance, speed, tuning, ADSR, etc. You can also vary the modulation bus levels to see their influence.

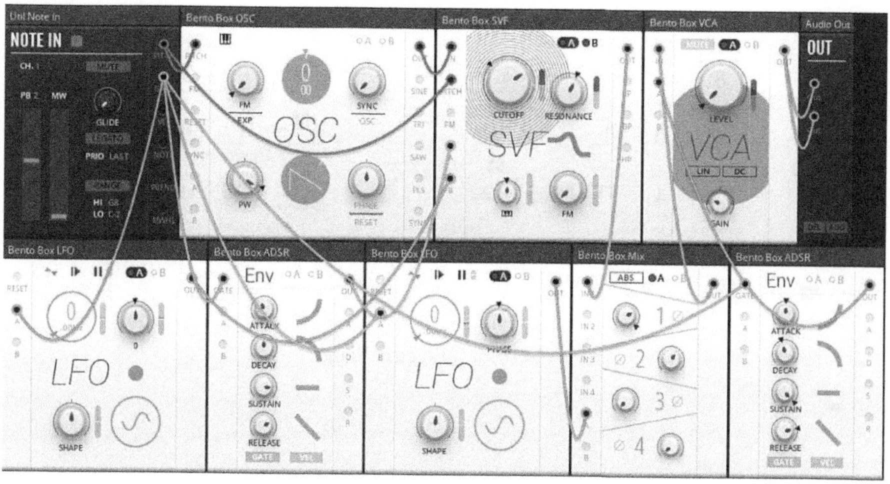

Figure 2.26. *The different Reaktor blocks in panel mode. For a color version of this figure, see www.iste.co.uk/reveillac/synthesizers2.zip*

2.3. Exercise 6 – multiple oscillators

With several oscillators, it resembles the characteristics of most synthesizers on the market, both hardware and software.

Typically, their VCOs can create multiple waveforms over different frequency ranges (different octaves), which can be tuned independently, generating complex sounds.

By detuning the oscillators, chords can be created by pressing a single key on the keyboard: a privilege for a monophonic synthesizer!

2.3.1. *Minimoog (or clones)*

The Minimoog has three oscillators covering six very different signal shapes (triangular, sawtooth triangular, square, wide rectangular and narrow rectangular) on octaves 2' to 32'.

Here, we have two patches for the Minimoog using the three oscillators.

The first, in Figure 2.27, is a sound imitating an electric piano and the second, in Figure 2.28, is a very long bass sound, similar to that made by the Minimoog.

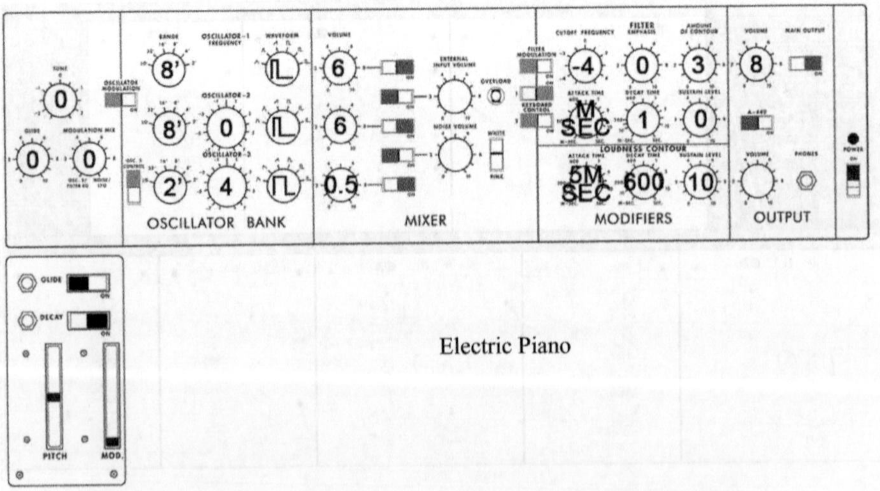

Figure 2.27. *Electric piano with a Minimoog. For a color version of this figure, see www.iste.co.uk/reveillac/synthesizers2.zip*

Instructions and comments are as follows:

– position the wheels, switches and dials as shown in Figure 2.27;

– PITCH wheel: 0;

– MOD wheel: 0;

– TUNE: 0;

– GLIDE: OFF;

– DECAY: O;

– OSCILLATOR MODULATION: OFF;

– OSCILLATOR 1: RANGE = 8' – WAVEFORM = rectangle (very narrow);

– OSCILLATOR 1: VOLUME: 6;

– OSCILLATOR 2: RANGE = 8' – WAVEFORM = rectangle (very narrow);

– OSCILLATOR 2: FREQUENCY = 0;

– OSCILLATOR 2: VOLUME: 6;

– OSC.3 CONTROL: OFF;

– OSCILLATOR 3: RANGE = 2' – WAVEFORM = square;

– OSCILLATOR: FREQUENCY = 4;

– OSCILLATOR 3: VOLUME: 0.5;

– MIXER switches: ON; OFF; ON; OFF; ON;

– FILTER MODULATION: OFF;

– KEYBOARD CONTROL 1: ON;

– KEYBOARD CONTROL 2: OFF;

– FILTER: CUTOFF = – 4; EMPHASIS = 0; CONTOUR = ;

– envelope 1 (FILTER): ATTACK TIME = M-Sec; DECAY TIME = 1 s; SUSTAIN LEVEL = 0;

– envelope 2 (MODIFIERS): ATTACK TIME = 5 M-Sec; DECAY TIME = 600 ms; SUSTAIN LEVEL = 10;

– A440: OFF;

– OUTPUT VOLUME: 8;

– MIX OUTPUT: ON.

Let us move on to a more iconic sound that the Minimoog makes with the Long Sub Bass patch.

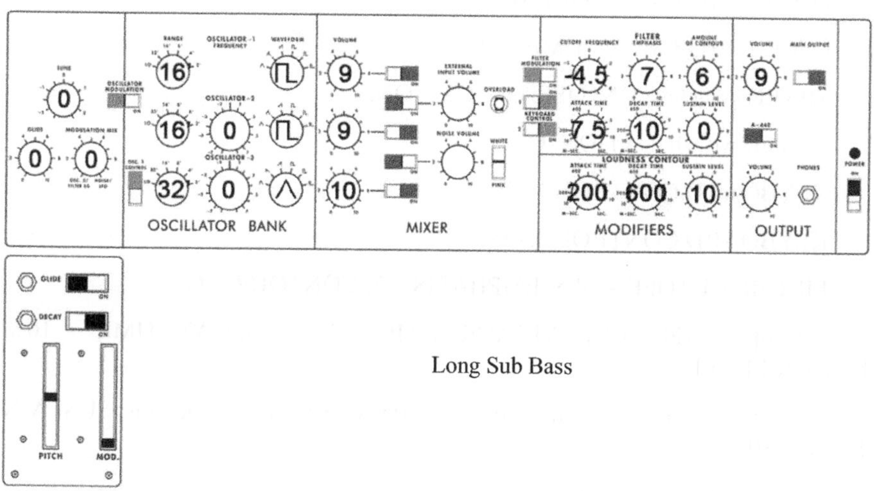

Figure 2.28. *An iconic Minimoog bass sound. For a color version of this figure, see www.iste.co.uk/reveillac/synthesizers2.zip*

Instructions and comments are as follows:

– position the wheels, switches and dials as shown in Figure 2.2;

– PITCH wheel: 0;

– MOD wheel: 0;

– TUNE: 0;

– GLIDE: OFF;

– DECAY: ON;

– OSCILLATOR MODULATION: OFF;

– OSCILLATOR 1: RANGE = 16' – WAVEFORM = square;

– OSCILLATOR 1: VOLUME: 9;

– OSCILLATOR 2: RANGE = 16' – WAVEFORM = square;

– OSCILLATOR 2: FREQUENCY = 0;

– OSCILLATOR 2: VOLUME: 9;

– OSC.3 CONTROL: OFF;

– OSCILLATOR 3: RANGE = 32' – WAVEFORM = triangle;

– OSCILLATOR 3: FREQUENCY = 0;

– OSCILLATOR 3: VOLUME: 10;

– MIXER switches: ON; OFF; ON; OFF; ON;

– FILTER MODULATION: OFF;

– KEYBOARD CONTROL 1: ON;

– KEYBOARD CONTROL 2: ON;

– FILTER: CUTOFF = –4.5; EMPHASIS = 7; CONTOUR = 6;

– envelope 1 (FILTER): ATTACK TIME = 7.5 s; DECAY TIME = 10 s; SUSTAIN LEVEL = 0;

– envelope 2 (modifiers): attack time = 200 ms; decay time = 600 ms; SUSTAIN LEVEL = 10;

– OUTPUT VOLUME: 9;

– MIX OUTPUT: ON;

– to adjust the sound rendering to your taste, you can slightly adjust the filter's cutoff frequency and emphasis.

2.3.2. *Novation Bass Station II*

The patch chosen here is a Moog-type bass sound; it uses the two oscillators, the sub-oscillator, the two LFOs and the two envelope generators of the Bass Station II.

The richness of the intense sound shows a controlled use of the oscillators. The temporal progression of the sound when pressing a key on the keyboard is interesting and highlights the use of two very different envelopes.

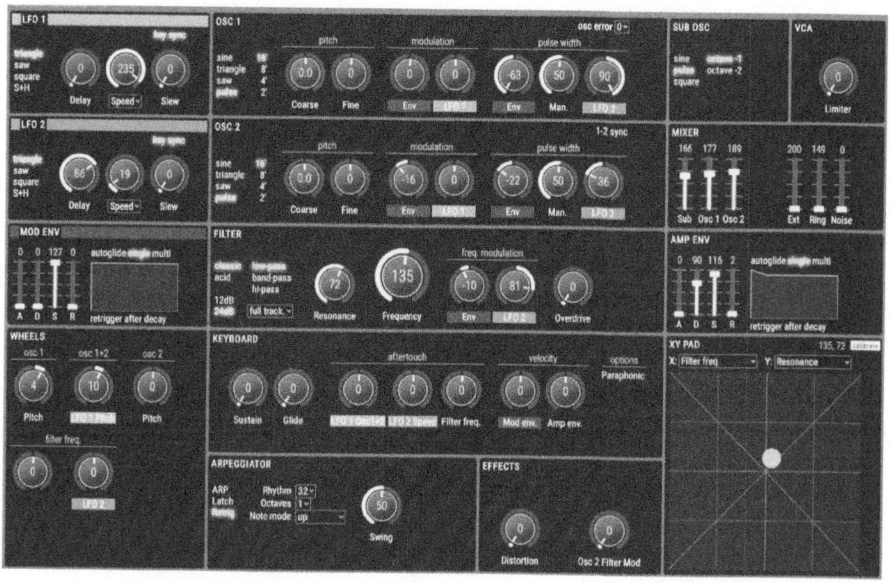

Figure 2.29. *A very typical Moog bass sound. For a color version of this figure, see www.iste.co.uk/reveillac/synthesizers2.zip*

Instructions and comments are as follows:

– set the values of the different commands as indicated below;

– LFO 1: SPEED = 235;

– LFO 2: DELAY = 86, SPEED = 19;

– OSC 1: PULSE 16’ – ENV = –63, MAN. = 50, LFO 2 = 90;

– OSC 2: PULSE 16’ – ENV = –22, MAN. = 50, LFO2 = 36;

– SUB OSC: PULSE, OCTAVE-1;

– MOD ENV: S = 127;

– AMP ENV: D = 90, S = 116, SINGLE;

– FILTER: CLASSIC, LOW-PASS, 24 dB, FULL TRACK;

– FILTER: RESONANCE = 72, FREQUENCY = 135, FREQ. MODULATION – ENV = –10, LFO 2 = 81;

– MIXER: SUB = 166, OSC 1 = 177, OSC 2 = 189;

– WHEELS: OSC 1 = 4, OSC 1 + 2 = 10;

– ARPEGGIATOR: default;

– all other parameters: 0;

– remember to set the output volume to the correct value.

2.3.3. *Max/MSP*

For this exercise, we will mix the signals of three oscillators with multiple separately tunable waveforms (sine, sawtooth, triangular, rectangular), as can be found in an ARP 2600 type synthesizer, Minimoog or Arturia MatrixBrute.

The Max/MSP patcher in Figure 2.30 uses the `VCOChoice` abstraction to work correctly.

The abstraction combines a selector, allowing you to choose one of the four oscillators generating different waveforms (sinusoidal, sawtooth, triangular and rectangular) and the device for initializing the sliders that tune each oscillator.

Instructions and comments are as follows:

– let us start by creating the `VCOChoice` abstraction (see Figure 2.30);

– it uses a setup like one already encountered;

– the `mtof`, `flonum`, `cycle~`, `phasor~`, `triangle~`, `rect~` and `selector~ 4` objects will provide the signal, coming from the MIDI keyboard, for each oscillator with four different waveforms, one of which will be chosen by the user;

– the `loadbang` and `message 63` objects will initialize the `slider` object at their central positions, 0 (right tuning: A440);

– the four `inlet` objects will, respectively, receive the user's choice of waveform via a drop-down menu (umenu), the note coming from the MIDI keyboard, the pitch of the note chosen by the user via the tuning `slider` object;

– the two `outlet` objects will, respectively, provide the signal following the chosen waveform and the initialization of the `slider` objects for adjusting the tuning of the oscillators;

– make all connections as shown in Figure 2.30;

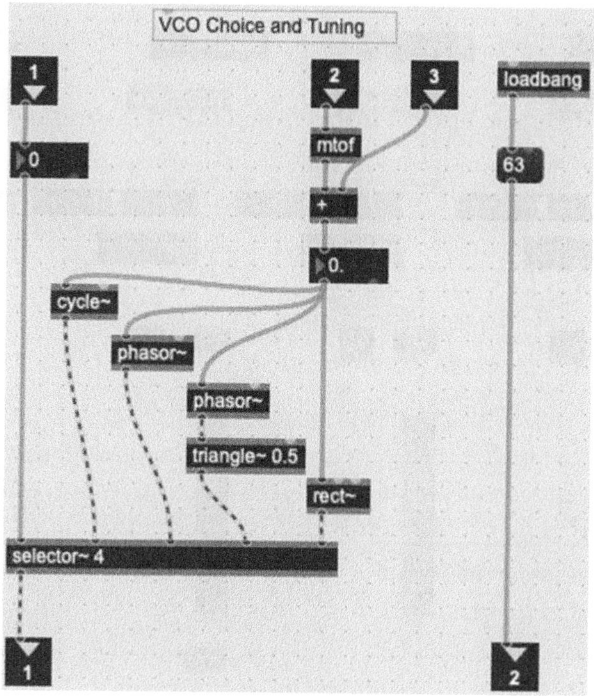

Figure 2.30. *The* `VCOchoice` *abstraction used by the patcher*

– now, for the main patcher (Figure 2.31), create a new patcher;

– create a `notein` object to receive notes from the MIDI keyboard;

– create three umenu objects (`VCO 1`, `VCO 2`, `VCO 3`) with the options `None`, `Sinus`, `Sawtooth`, `Triangle` and `Rectangle`;

– create three `VCOChoice` abstraction objects as created previously;

– create three `slider Tune` objects with size `127.` with a minimum output value of –63. These objects will define the tuning of each of the three oscillators;

– create three `number` objects that will display the negative or positive value of the agreement;

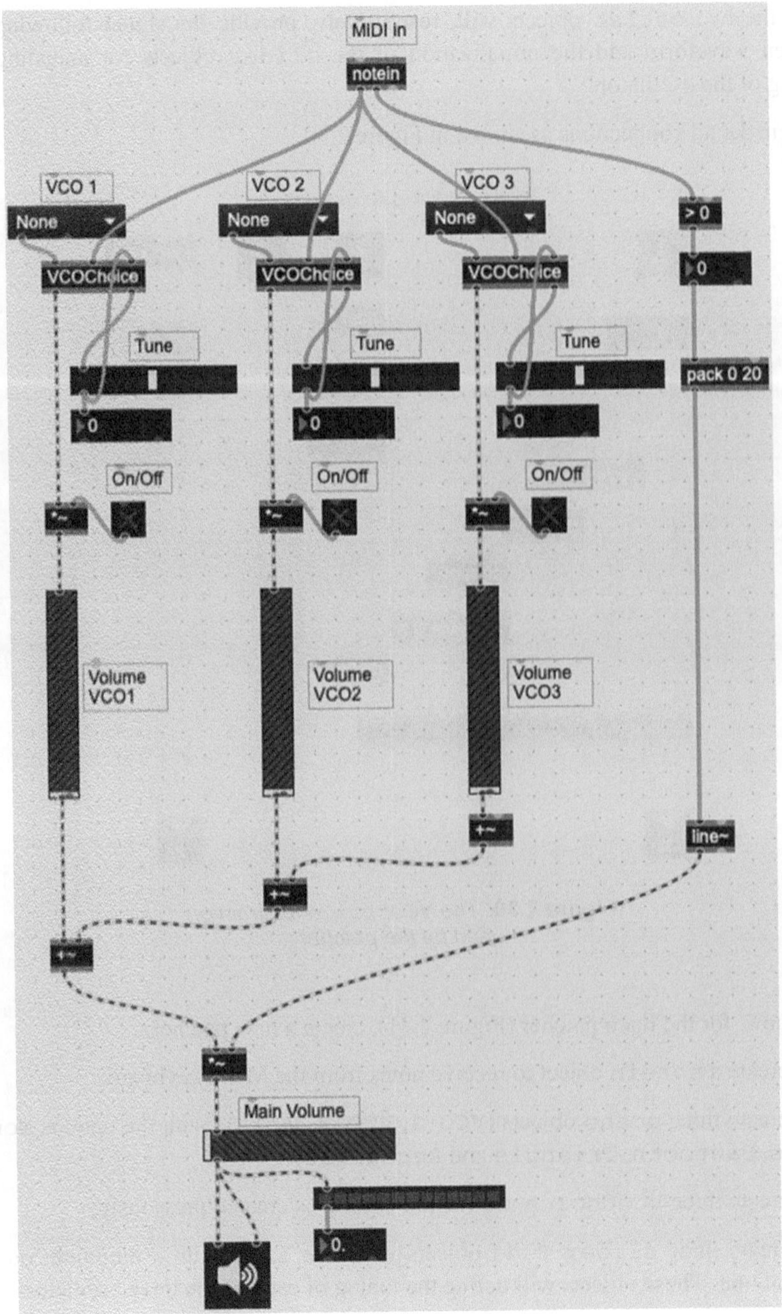

Figure 2.31. *The multi-oscillator patcher*

– for each oscillator, connect the left output of the `notein` object to the second input of the `VCOChoice` abstraction;

– for each oscillator, connect the output of the umenu object to the first input of the `VCOChoice` abstraction;

– for each oscillator, connect the second output of the `VCOChoice` abstraction to the `slider Tune` object;

– connect the output of each `slider Tune` object to the third input of the `VCOChoice` abstraction object;

– create three *~ objects and three `toggle` objects to activate the passage of the signal from each oscillator;

– create three `gain` objects to adjust the output volume of each oscillator;

– create three +~ objects to add and mix signals from the oscillators;

– for each oscillator, connect the first output of the `VCOChoice` abstraction to the left input of the object *~;

– connect the output of the *~ object to the input of the `gain` object;

– connect each `toggle` object to the right input of object *~ ;

– connect the output of each `gain` object to a +~ object;

– set up master volume management as shown in Figure 2.31;

– create a `> 0` object (test for the presence of a velocity), a `number` object (display velocity), a `pack 0 20` object and a `line~` object (20 ms increase), which makes it possible for a note to have a non-continuous signal when pressing a key on the MIDI keyboard;

– connect them and then the input of `> 0` to the central output of `notein` (velocity);

– connect the output of the `line~` object to the right input of the *~ object to apply the velocity increase to the oscillator signal;

– for the patcher to work, do not forget to choose a waveform, activate the signal pass for at least one oscillator, adjust the volume of each oscillator and the main volume;

– try slightly detuning the oscillators together, mixing different waveforms to give thickness to the sound. During a slight variance, we can hear the oscillators' beat frequency (see Volume 1, section 1.7.9).

2.3.4. *Pure Data*

For this exercise, we are going to create a new Pure Data patch that uses some elements already encountered and also adds new ones, such as `VU-meter`, `r` (receive) and `s` (send) objects.

Two oscillators will be available and the user can choose their waveforms, tuning and activation. The volume of each can be adjusted and will be visible on a VU meter.

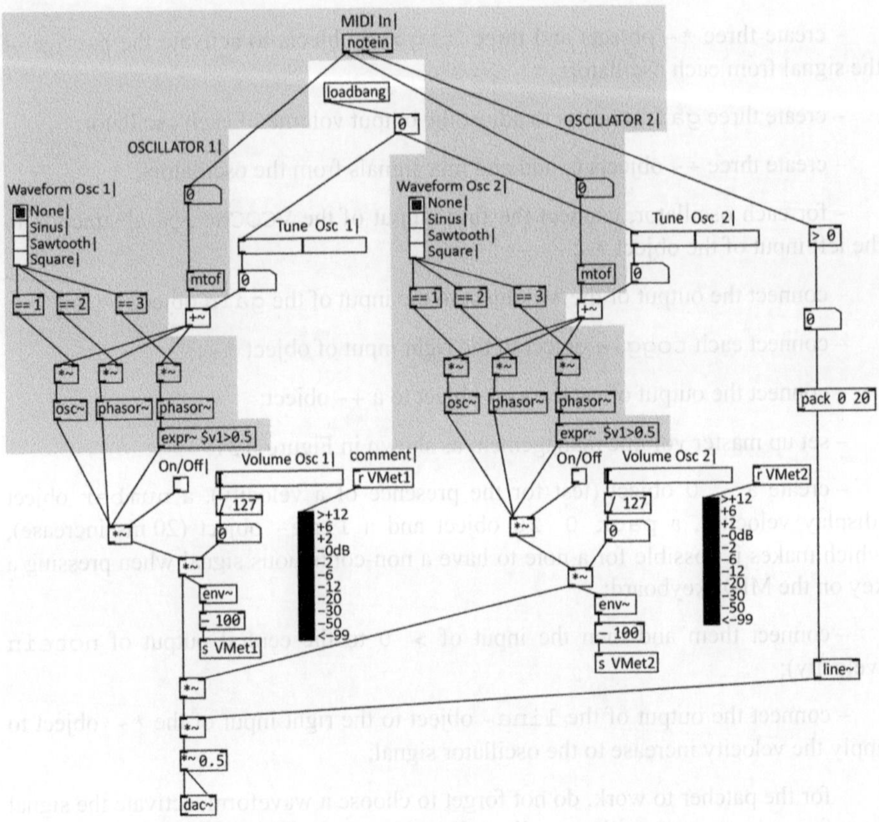

Figure 2.32. *Sharing two oscillators in Pure Data*

Instructions and comments are as follows:

– objects (on gray background, Figures 2.32 and 2.33) `notein`, `number`, `mtof`, `Vradio`, `== 1`, `== 2`, `== 3`, `*~`, `osc~`, `phasor~`, `expr~ $v1>0.5` have already been described and do not bring any new functionalities. They form a system of two oscillators controlled by a MIDI keyboard;

– create and connect the different objects as indicated (Figure 2.32);

– create a `loadbang` object (Figure 2.33) connected to the input of a `0` object to initialize `slider` objects that control the volume in the central position (0) when opening the Pure Data file;

– create two `slider` objects to control the tuning of each oscillator. They are followed by two `number` objects to control the value;

– these two `slider` objects have a linear range of values between –63 and 64;

– connect the output of object `0` to the two inputs of the two `slider` objects;

– create two `+~` objects that will add or subtract the value of the `slider` object to modify the oscillator tuning;

– connect the output of each `number` object to the right input of the `+~` objects located behind each `mtof` object;

– connect the output of each `mtof` object to the right input of the `+~` object;

– the `> 0`, `number`, `pack 0 20` and `line~` objects placed to the right have already been seen in previous exercises; they prevent a continuous sound when pressing a key on a keyboard;

– the `slider Volume Osc 1`, `slider Volume Osc 2`, `/ 127`, `number` and `*~` objects, as we have seen previously, control the volume of each oscillator;

– the three `env~`, `- 100` and `s Vmet` (1-2) objects placed behind the object `*~`, send the value in decibels to the object `r Vmet` (1-2). The `env~` object gives the volume in decibels. The object `- 100` is mandatory; by default, the number of decibels always increases by 100;

– the two `VU meter` objects ensure the decibels are displayed (left input);

– the object `+~` ensures the mixing of the signals from the two oscillators;

– the two `toggle On/Off` objects, each followed by a `*~` object, activate or deactivate the passage of the sound signal from an oscillator to the output;

– the three objects that follow, `*~`, `*~ 0.5` and `dac~`, conventionally ensure the output of the audio signal, with a power equal to 50% by default;

– remember to choose a waveform, activate at least one oscillator and adjust the volume; otherwise, no sound will be produced;

– try the different settings and options, and play with the detuning of the oscillators, the volume of each, and their waveforms; you will obtain very different sound reproductions. Mixed sounds from multiple oscillators always tend to thicken the sound signal.

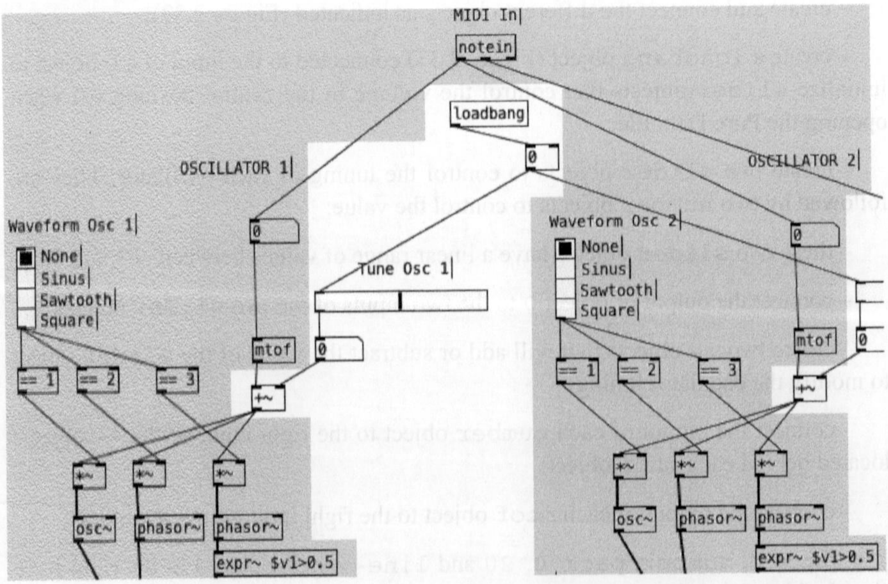

Figure 2.33. *Details of the Pure Data objects forming the two oscillators and their commands*

2.3.5. *VCV Rack*

The exercise presented below brings together three oscillators around a mixer associated with a cutoff module. We find two envelope generators, the filter and the VCA.

The layout of the modules has been modified compared to the previous exercise to have more readable connections.

Instructions and comments are as follows:

– take the rack created in the previous exercise (Figure 2.22);

– remove the two LFO modules and the SCOPE module;

– remove the cables connected to the VCA MIX module;

– add two VCO modules and two OCT modules;

– connect the V/OCT output of the MIDI-CV module to the two V/OCT inputs of the new OCT modules;

– connect the two outputs of each OCT module to the V/OCT inputs of each new VCO module;

– choose an output (SIN, TRI, SAW, or SQR) of the three VCO modules and connect them to three IN inputs of the MUTES module. This module has push buttons on each of its inputs that cut off the incoming signal;

– connect the three outputs of the MUTES module corresponding to the previous inputs to the three inputs (IN 1, IN 2 and IN 3) of the VCA MIX module;

– connect the MIX output of the VCA MIX module to the IN input of the VCF module;

– try the possible combinations of waveforms on the VCOs by modifying the tuning (FREQ) and the octave on the OCT modules. You can obtain richer sounds when reproducing a sound using multiple oscillators.

Figure 2.34. *Several oscillators with a VCV Rack. For a color version of this figure, see www.iste.co.uk/reveillac/synthesizers2.zip*

2.3.6. *Reaktor*

Compared to the previous one, this new Reaktor rack has two additional oscillators and a mixing module to mix and combine the different signals generated by the oscillators.

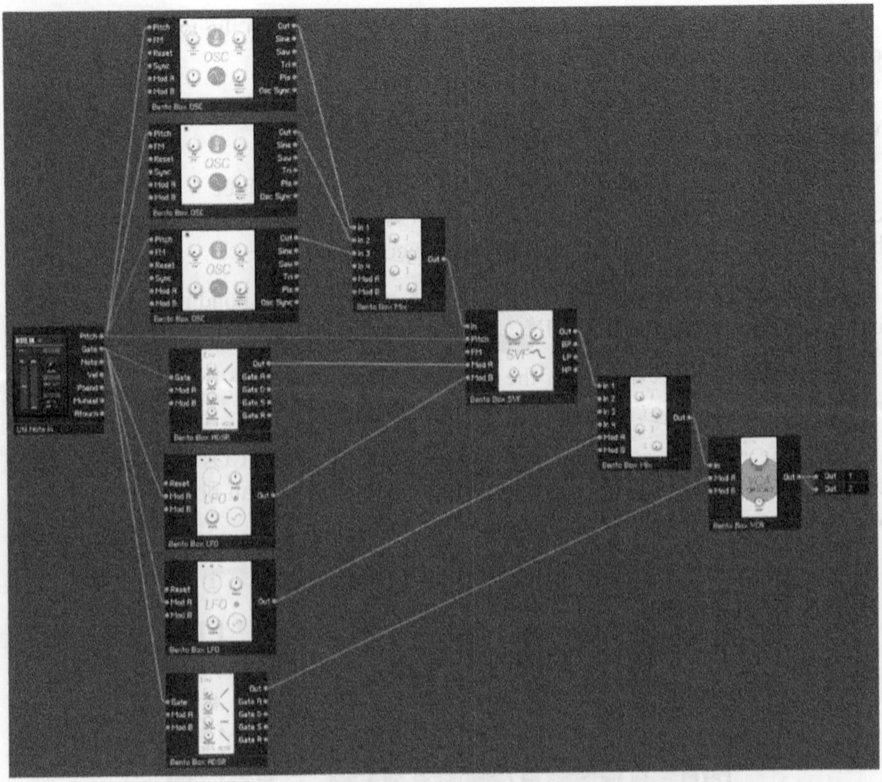

Figure 2.35. *Three oscillators for a Reaktor structure. For a color version of this figure, see www.iste.co.uk/reveillac/synthesizers2.zip*

Instructions and comments are as follows:

– repeat the previous exercise;

– remove the connection between the OUT output of the OSC block and the IN input of the SVF block;

– add a MIX-Bento Mix block and duplicate the OSC block twice;

– arrange the blocks as shown in Figure 2.35;

– connect everything as described in Table 2.4.

Figure 2.36 shows the new rack created in panel mode; the modules are positioned in three rows for better readability of this new configuration.

Block	Out	Block	In	Comments
NOTE IN	Pitch	OSC 1	Pitch	Sends the MIDI note from the keyboard to oscillator 1
NOTE IN	Pitch	OSC 2	Pitch	Sends the MIDI note from the keyboard to oscillator 2
NOTE IN	Pitch	OSC 3	Pitch	Sends the MIDI note from the keyboard to oscillator 3
NOTE IN	Gate	ENV 1	Gate	Triggers envelope 1
OSC 1	Out	MIX (new)	In 1	Sends the signal from oscillator 1 to the mixer
OSC 2	Out	MIX (new)	In 2	Sends the signal from oscillator 2 to the mixer
OSC 3	Out	MIX (new)	In 3	Sends the signal from oscillator 3 to the mixer
MIX (new)	Out	SVF	In	Sends the mixed signal to the SVF filter

Table 2.4. *New connections to make for our multi-oscillator rack*

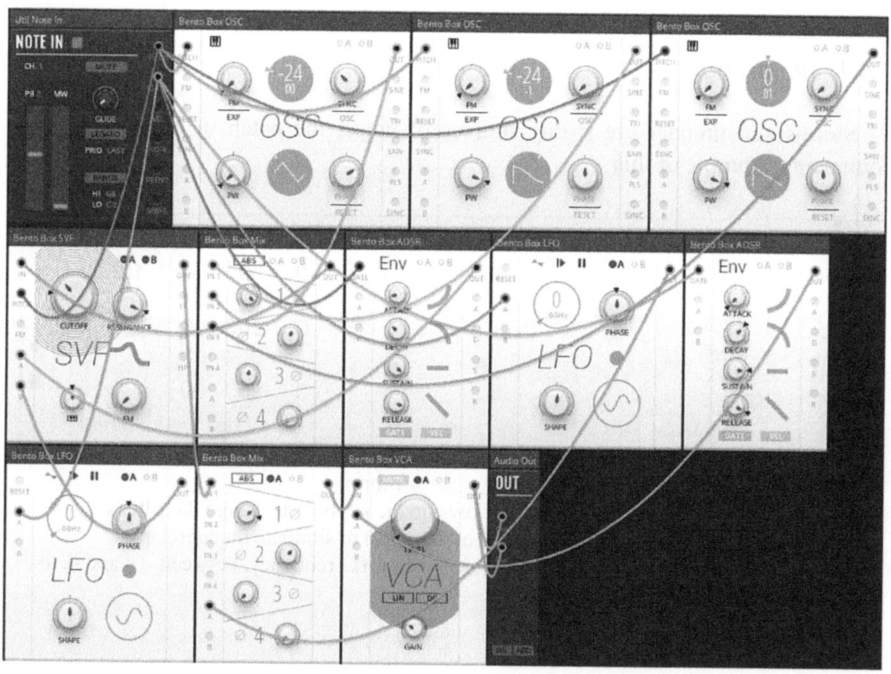

Figure 2.36. *Three oscillators for Reaktor in the panel mode. For a color version of this figure, see www.iste.co.uk/reveillac/synthesizers2.zip*

2.4. Exercise 7 – noise generator

Suppose oscillators are the critical elements of subtractive synthesis within synthesizers. In that case, we must not neglect the noise generator. This somewhat unique oscillator generates a random set of waveforms over a defined frequency range whereby all pitches are present (see Volume 1, section 1.8). Noise is a phenomenon fully integrated into our daily sound environment.

The creation of specific sound effects such as sea surf, wind, rain, storms, a steam locomotive, noises emitted by various machines and mechanisms (fan, blower, turbine, compressor, etc.), or even the sound of some wind or percussion instruments involves a noise generator to synthesize them.

Noise is characterized by its color, white, pink or blue, which defines the frequency range over which it is most present (see Volume 1, sections 1.8.1–1.8.7).

In the following exercises, we will develop some particular sounds for which the presence of noise is significant.

2.4.1. *Minimoog*

Storms and thunder are strong sources of noise; the patch in Figure 2.37 is a convincing example of this.

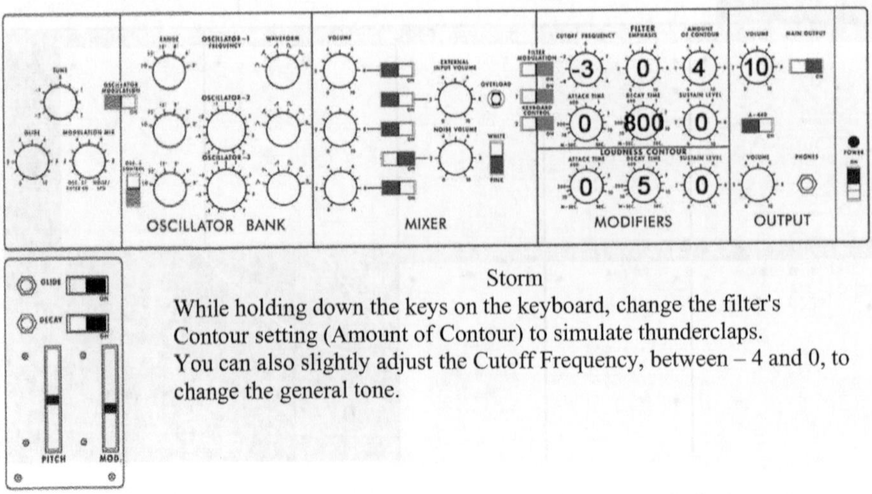

Figure 2.37. *A patch to simulate storms and wind. For a color version of this figure, see www.iste.co.uk/reveillac/synthesizers2.zip*

Instructions and comments are as follows:

– position the wheels, switches and dials as shown in Figure 2.36;

– PITCH wheel: 0;

– MOD wheel: 30%;

– TUNE: 0;

– GLIDE: ON;

– DECAY: ON;

– OSCILLATOR MODULATION: OFF;

– OSC.3 CONTROL: ON;

– MIXER switches: OFF; OFF; OFF; ON; OFF;

– NOISE: PINK;

– FILTER MODULATION: ON;

– KEYBOARD CONTROL 1: ON;

– KEYBOARD CONTROL 2: ON;

– FILTER: CUTOFF = –3 EMPHASIS = 0; CONTOUR = 4;

– envelope 1 (FILTER): ATTACK TIME = 0; DECAY TIME = 800; SUSTAIN LEVEL = 0;

– envelope 2 (MODIFIERS): ATTACK TIME = 0; DECAY TIME = 5; SUSTAIN LEVEL = 0;

– A440: OFF;

– OUTPUT VOLUME: 10;

– MIX OUTPUT: ON.

In Figure 2.38, you will find a patch simulating the sound of a flute, a touch of white noise adds the musician's breath.

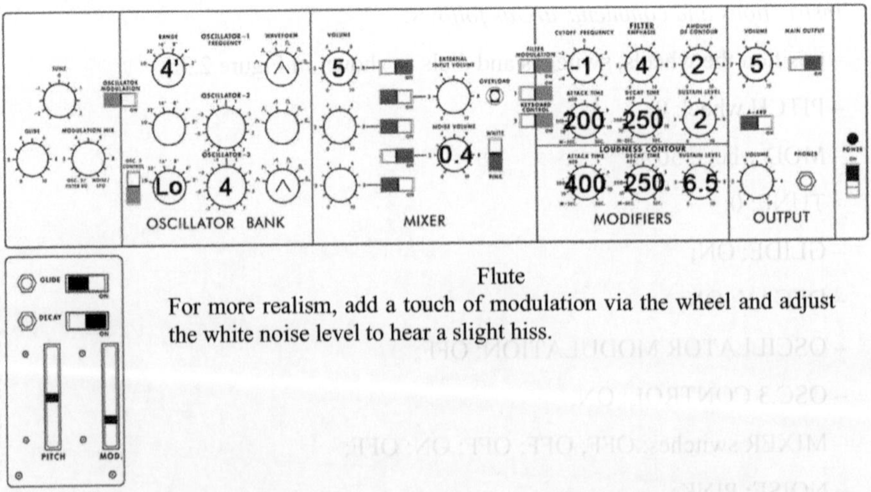

Figure 2.38. *The sound of a flute. For a color version of this figure, see www.iste.co.uk/reveillac/synthesizers2.zip*

Instructions and comments are as follows:

– position the wheels, switches and dials as shown in Figure 2.38;

– PITCH wheel: 0;

– MOD wheel: 0–10%;

– TUNE: 0;

– GLIDE: OFF;

– DECAY: ON;

– MODULATION MIX: 0;

– OSCILLATOR MODULATION: OFF;

– OSCILLATOR 1: RANGE = 4' – WAVEFORM = triangle;

– OSCILLATOR 1: VOLUME: 5;

– OSC.3 CONTROL: ON;

– OSCILLATOR 3: RANGE = LO – WAVEFORM = triangle;

– OSCILLATOR 3: FREQUENCY = 4;

– MIXER switches: ON; OFF; OFF; ON; OFF;

– NOISE VOLUME: 0.4;

– NOISE: PINK;

– FILTER MODULATION: ON;

– KEYBOARD CONTROL 1: ON;

– KEYBOARD CONTROL 2: ON;

– FILTER: CUTOFF = –1; EMPHASIS = 4; CONTOUR = 2;

– envelope 1 (FILTER): ATTACK TIME = 200; DECAY TIME = 250; SUSTAIN LEVEL = 2;

– envelope 2 (MODIFIERS): ATTACK TIME = 400; DECAY TIME = 250; SUSTAIN LEVEL = 6.5;

– A440: OFF;

– OUTPUT VOLUME: 5;

– VOICE OUTPUT: ON.

2.4.2. *Behringer 2600 (ARP 2600)*

The patch in Figure 2.39 combines the settings of the Behringer 2600 (ARP 2600) to produce the sound of an explosion.

It fully implements the noise generator associated with the synthesizer of oscillator 3, which generates a low-frequency signal.

Instructions and comments are as follows:

– copy the patch from Figure 2.39;

– for more realism, you can add a little reverb on each channel before the signal output;

– modifications to the attack time and the release time can also be made.

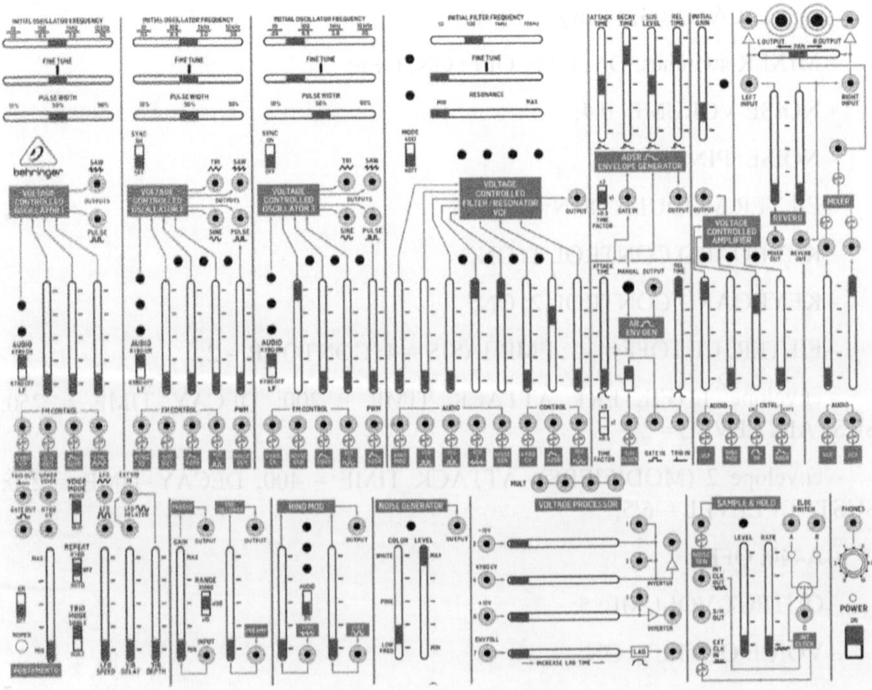

Figure 2.39. *Patch explosion on the Behringer 2600. For a color version of this figure, see www.iste.co.uk/reveillac/synthesizers2.zip*

2.4.3. *Novation Bass Station II*

The patch in Figure 2.40 simulates the sound of a steam locomotive. Press a key in the lower range of the keyboard (octave 0) and gradually increase in speed to hear the train depart.

By modifying the release time of the VCA envelope (by default 28), you can improve the realism of the rendered sound, particularly if the train decelerates and enters the station.

Instructions and comments are as follows:

– set the values of the different commands as indicated below;

– SUB OSC: SINE, OCTAVE-1;

– AMP ENV: A = 17; D = 79, S = 100, R = 28, SINGLE;

– FILTER: CLASSIC, HI-PASS, 12 dB, FULL TRACK;

– FILTER: RESONANCE = 63, FREQUENCY = 90, FREQ. MODULATION – ENV = 0, LFO 2 = 20, OVERDRIVE = 55;

– MIXER: NOISE = 255;

– ARPEGGIATOR: default;

– All other parameters: 0;

– remember to set the output volume to the correct value.

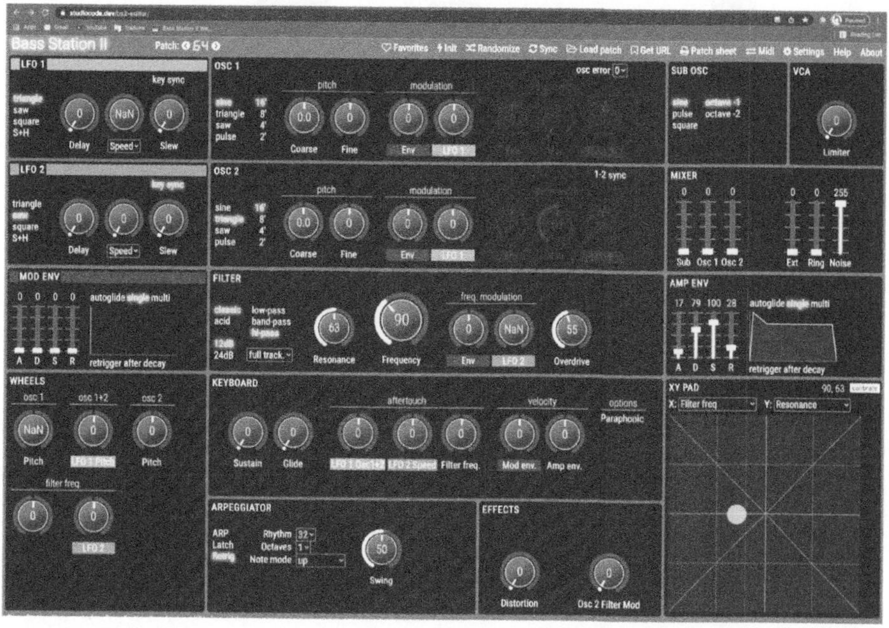

Figure 2.40. *The steam locomotive patch using the Bass Station II noise generator. For a color version of this figure, see www.iste.co.uk/reveillac/synthesizers2.zip*

2.4.4. *Max/MSP*

In this exercise, an oscillator, whose frequency and waveform can be specified (sinusoidal, sawtooth, triangular, rectangular), modulates a signal composed of white noise transmitted through a band-pass filter, whose parameters can be adjusted.

This simulator allows the synthesis of several characteristic noises, such as that of a helicopter, a small single-engine plane, rain, a steam train, etc.

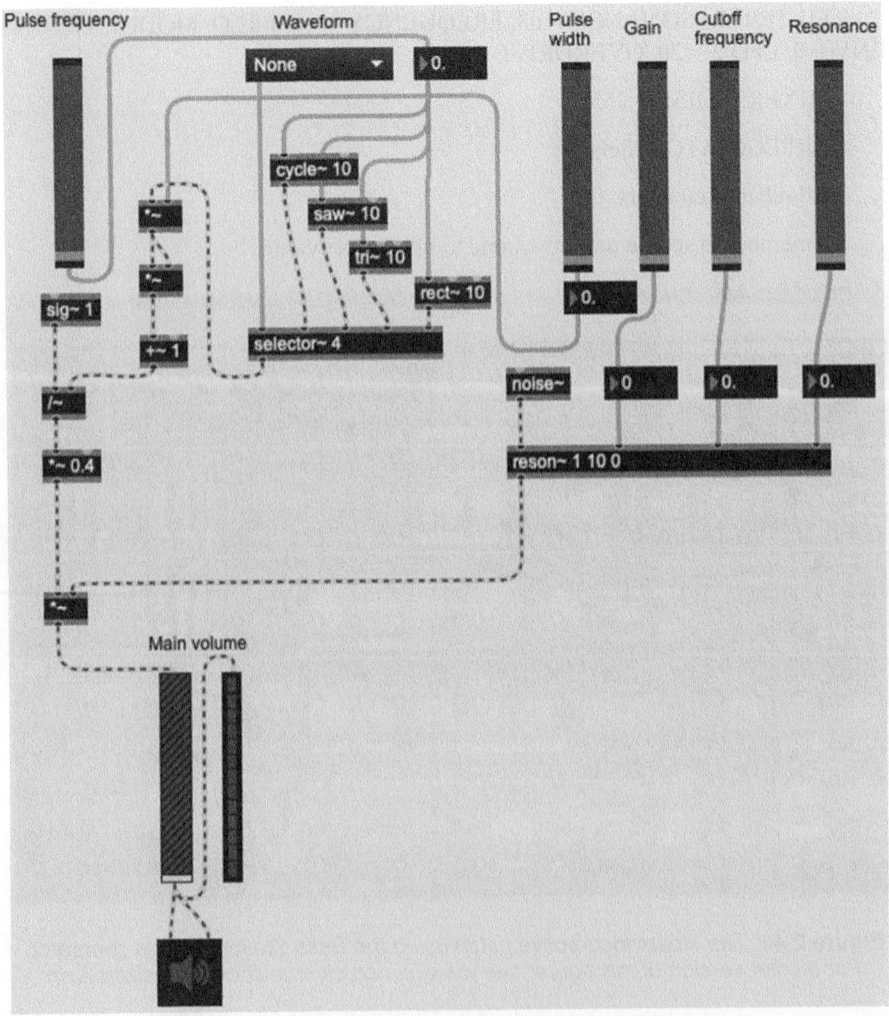

Figure 2.41. *Noise simulator with the Max/MSP*

Instructions and comments are as follows:

– let us start by creating the user control part;

– create five `slider` objects and four `flonum` objects to define user commands;

– connect the output of each `slider` object at the input of each `flonum` object;

– the slider Pulse Frequency object has a size of 50 and a minimum output decimal value equal to 0;

– the slider Pulse Width object has a size of 10 and a minimum output decimal value of 0;

– the slider Gain object has a size of 9 and a minimum output decimal value of 1;

– the slider Cutoff Frequency object has a size of 991 and a minimum output value equal to 10;

– the slider Resonance object has a size of 1 and a minimum output value of 0;

– create a umenu object, a drop-down menu with five items (None, Sinusoidal, Sawtooth, Triangular, Rectangular);

– create a cycle~10 object for the sinusoidal oscillator, which has a frequency of 10 Hz by default;

– create a saw~ 10 object for the sinusoidal oscillator, which has a frequency of 10 Hz by default;

– create a tri~ 10 object for the sinusoidal oscillator, which has a frequency of 10 Hz by default;

– create a rect~ 10 object for the sinusoidal oscillator, which has a frequency of 10 Hz by default;

– create a selector~ 4 object, which selects the choice made by the user in the drop-down menu;

– connect the output of the umenu object at the left input of the selector~ 4 object;

– connect the outputs of each oscillator to the remaining inputs of the selector~ 4 object;

– connect the output of the flonum object that follows the slider Pulse frequency object at each oscillator input (cycle~, saw~, tri~, rect~);

– let us continue with the modulator part;

– create a sig~ 1 object, which converts a number, here 1 by default, into an audio signal;

– create a /~ object, which attenuates and inverts the audio signal from the chosen oscillator;

– connect the output of the object sig~ 1 to the left input of object /~;

– create a *~ 0.4 object, which attenuates the signal from the modulator;

– connect the output of the object /~ to the left input of object *~ 0.4;

– create a *~ object, which combines the noise coming from the filter (carrier) with the signal from the modulator;

– connect the output of the object *~ 0.4 to the left input of object *~;

– create two *~ objects and connect them; then create a +~ 1 object and connect it to the output of the previous *~ object. These three objects shape (pulse) the signal coming from the chosen oscillator;

– connect the output of the flonum object that follows the slider Pulse width object with the input of the first object *~.

– connect the output of the object +~ 1 with the right input of the /~ object;

– let us move on to the noise generator part (carrier);

– create a reson~ 1 10 0 object to define a bandpass filter whose gain, cutoff frequency and resonance will default to 1,10 and 0;

– create a noise~ object to generate white noise;

– connect the output of the noise~ object to the left input of the reson~ 1 10 0 object to filter the signal;

– connect the outputs of the three flonum objects attached to each slider, Gain, Cutoff frequency and Resonance object to the remaining inputs of the reson~ 1 10 0 object in this order;

– finally, let us take care of the main volume control part;

– connect the output of the reson~ 1 10 0 object to the right input of the *~ object created in the modulator part of this exercise;

– create a gain object for master volume control;

– create an ezdac~ object for audio output;

– create a meter~ object to control the master volume;

– connect the output of the *~ object to the input of the gain object;

– connect the output of the gain object to both inputs of the ezdac~ object;

– connect the output of the gain object to the input of the meter~ object;

– remember to adjust the master volume and choose a waveform for a modulated signal; otherwise, you will only get noise. Use the different controls to shape the sound that suits you.

Table 2.5 presents some simulator configurations.

Sound effect	Pulse freq.	Waveform	Pulse width	Gain	Cutoff freq.	Reson.	Main vol.
Helicopter	8	Sinusoidal	3	6	300	0.95	70%
Single-engine plane	15	Sinusoidal	6.5	9	190	0.5	70%
Locomotive	7.5	Triangular	5	. 3	400	0.5	80%
Rain	0	Rectangular	5.5	10	115	0.5	80%

Table 2.5. *Some miscellaneous sound effects*

2.4.5. *Pure Data*

This new Pure Data patch will allow you to explore the harmonic variations of white noise through a band-pass filter. You will also be able to listen to the influence of even and odd harmonics when they are present or absent.

Instructions and comments are as follows:

– create a `loadbang` object, which will load default values when opening the Pure Data patch;

– create three objects with values of 1,000 for the center frequency, 10 for bandwidth and 64 for the volume;

– create three `slider` objects, `Center frequency`, `Bandwidth` and `Volume`;

– the linear value ranges for each `slider` object will be 0–2,000, 0–50 and 0–127;

– connect the output of the `loadbang` object to the input of each object containing the values (1,000, 10 and 64);

– connect the output of each object containing these values to the input of each `slider` object;

– create two `number` objects, which control the center frequency and bandwidth values;

– connect the outputs of both `slider` objects, `center frequency` and `bandwidth` to the input of both `number` objects;

– create a `noise~` object, which will generate white noise;

– create a `vcf~ 1000 10` object, a bandpass filter initialized with a center frequency of 1,000 Hz, and a bandwidth equal to 10;

– connect the output of the `noise~` object to the left input of the `vcf~ 1000 10` object;

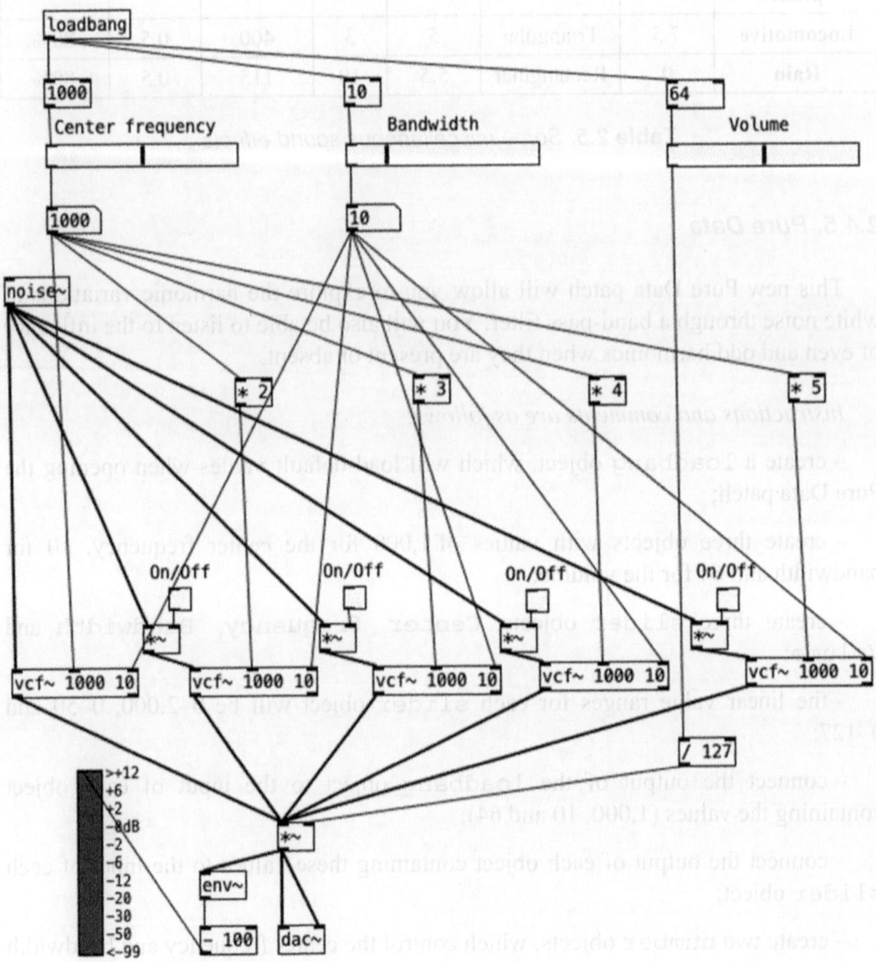

Figure 2.42. *White noise, bandpass filter and harmonics*

– create four objects `* 2`, `* 3`, `* 4` and `* 5`, which will be fundamental frequency multipliers to generate the even and odd harmonics h2, h3, h4 and h5;

– create four `toggle` objects, four `*~` objects and four `vcf~ 1000 10` objects;

– connect the output of each `toggle` object to the right input of each `*~` object;

– connect the output of each `*~` object to the left input of each `vcf~ 1000 10` object;

– connect the output of the `noise~` object to the left input of each of the four `*~` objects;

– connect the output of the `number` placed behind the `slider Center frequency` to the left inputs of the objects `* 2`, `* 3`, `* 4` and `* 5`;

– individually connect the outputs of the objects `* 2`, `* 3`, `* 4` and `* 5` to each middle input of the five `vcf~ 1000 10` objects;

– create a `/ 127` object to ensure volume control between 0 and 1 by dividing the value transmitted by the `slider Volume` object;

– connect the output of the `slider Volume` object to the left input of the object `/ 127`;

– create a `*~` object, an `env~` object, a `dac~` object, a `- 100` object in a `VU meter` object;

– connect the output of the `/ 127` object to the right input of the `*~` object;

– connect the output of the `*~` object to the input of the `env~` object, which converts a signal into decibels, and to both inputs of the `dac~` object;

– connect the output of the `env~` object to the left input of the `- 100` object to be compatible with the `VU meter` object;

– connect the output of the `- 100` object to the left input of the `VU Meter` object;

– when opening the patch, a sound signal should already be present. Test the different possible configurations by clicking the sliders and checkboxes.

2.5. To conclude classic synthesis

This chapter discussed the additional constituents within a synthesizer operating in subtractive synthesis: the envelope, LFO, multiple oscillators and noise generator.

The combination of the work presented in this chapter and the previous (Chapters 1 and 2) allows us to begin implementing a fully operational subtractive system.

However, other features are not always considered essential; we can discover these in the rest of this book.

3

Advanced Subtractive Synthesis

Many synthesizers include advanced features that differ from those we have already encountered. Ring modulation and sample and hold often complement the subtractive sound synthesis toolbox.

Moreover, various effects, generally placed behind the VCA, provide a finishing touch, producing an even more spectacular final sound.

As with Chapter 2, I will discuss exercises with hardware synthesizers (Minimoog, Novation Bass Station II, ARP 2600, Arturia MatrixBrute) while keeping the same software tools (Max/MSP, Pure Data, VCV Rack, and Reaktor).

Table 3.1 lists the hardware and software chosen for each exercise and the sections where you can find them.

Exercise	Hardware/software	Section
No. 8. Ring modulation	Behringer 2600 (ARP2600)	3.1.1
	Arturia MatrixBrute	3.1.2
	Novation Bass Station II	3.1.3
	Max/MSP	3.1.4
	Pure Data	3.1.5
	VCV Rack	3.1.6
	Reaktor	3.1.7
No. 9. Sample and hold	Behringer 2600 (ARP2600)	3.2.1
	Arturia MatrixBrute	3.2.2
	Max/MSP	3.2.3

Exercise	Hardware/software	Section
No. 9. Sample and hold	Pure Data	3.2.4
	VCV Rack	3.2.5
	Reaktor	3.2.6
No. 10. Sound effects – Reverb	Max/MSP	3.3.1.1
		3.3.1.2
	Pure Data	3.3.1.3
		3.3.1.4
	VCV Rack	3.3.1.5
	Reaktor	3.3.1.6
No. 11. Sound effects – Chorus	Max/MSP	3.3.2.1
	Pure Data	3.3.2.2
	VCV Rack	3.3.2.3
No. 12. Sound effects – Flanger	Max/MSP	3.3.3.1
		3.3.3.2
	Pure Data	3.3.3.3
	VCV Rack	3.3.3.4
No. 13. Sound effects – Phaser	Max/MSP	3.4.4.1
	Pure Data	3.4.4.2
		3.4.4.3
	VCV Rack	3.4.4.4

Table 3.1. *Summary of the different exercises*

3.1. Exercise 8 – ring modulation

Ring modulation is present on many synthesizers, providing metallic sounds often popular in electronic music. It is often confused with amplitude modulation, which we encountered in Chapter 2 during our work with the LFO (see section 2.2).

The fundamental difference between the two, apart from the sound reproduction, is the polarity of the modulating or modulator signal, which is unipolar for amplitude modulation and bipolar for ring modulation, as shown in this small Pure Data patch (see Figure 3.1). It displays the waveforms of a unipolar or bipolar modulator for the same sinusoidal signal with a frequency of 300 Hz and amplitude of 1.

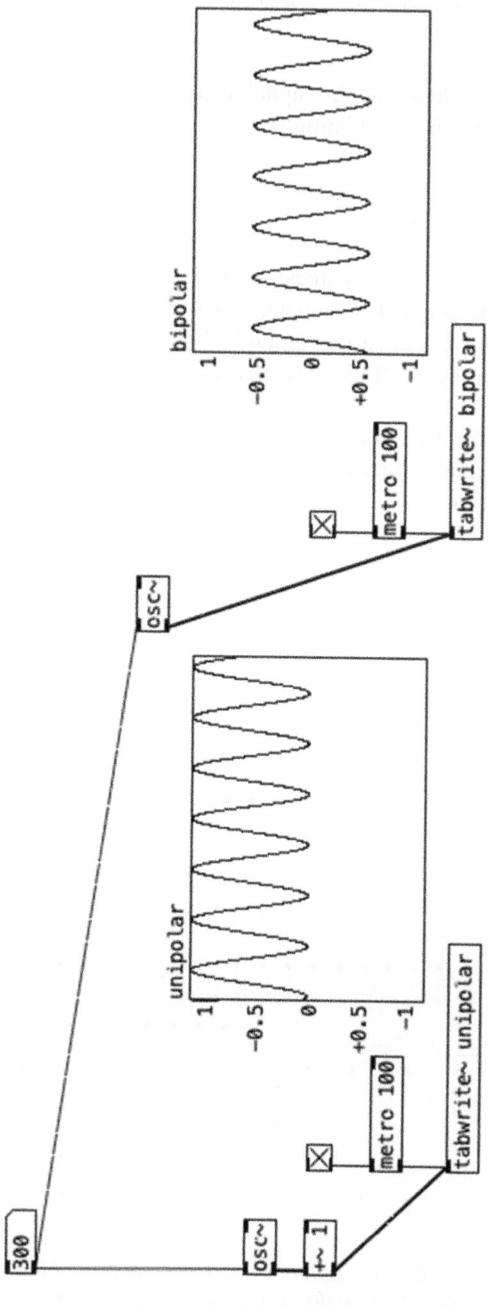

Figure 3.1. *Visualization of a unipolar and bipolar signal with Pure Data*

We can see that a unipolar signal has an amplitude between 0 and +1, while a bipolar signal has an amplitude between –0.5 and +0.5. The Pure Data object `+~ 1` changes the polarity.

For ring or amplitude modulation, the signal generated ultimately equals the product of the carrier signal by the modulating signal.

3.1.1. *Behringer 2600 (ARP 2600)*

The ARP 2600 was one of the first synthesizers to feature a ring modulator. The patch below comes from the original documentation supplied with the device, the famous "ARP 2600 Patch Book" containing around a hundred. This one (No. 7), the Tubular Chimes patch, is part of the "Basic Instruments" family.

It uses two oscillators; the first generates the carrier frequency and the second generates the modulating signal. Everything is sent to the filter and then the VCA, both shaped by their envelopes.

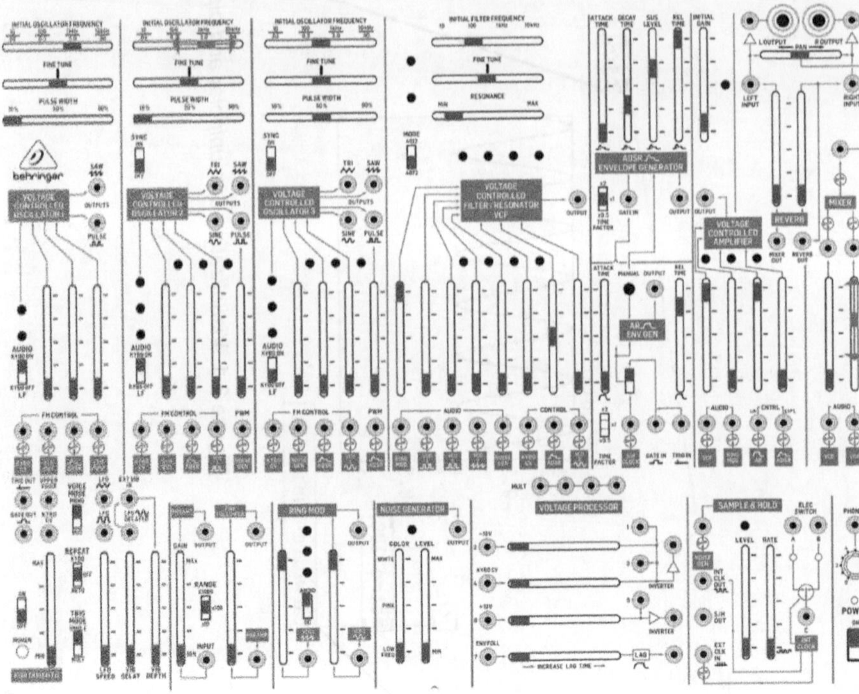

Figure 3.2. *Tubular Chimes patch. For a color version of this figure, see www.iste.co.uk/reveillac/synthesizers2.zip*

Instructions and comments are as follows:

– place the sliders and switches as shown on the patch in Figure 3.2;

– a little reverb can be added to make the sound more realistic;

– by very slightly altering the frequency of the second oscillator, you can refine the tone and modify the sound.

3.1.2. *Arturia MatrixBrute*

The MatrixBrute by Arturia is a large digitally-controlled analog synthesizer with many possibilities. It is a machine that we can use to exploit all aspects of subtractive synthesis.

Getting started may seem complex but with good knowledge of subtractive synthesis, once you understand the ergonomics of its human–machine interface, you will very quickly find your bearings.

The patch below generates a chime sound with two filtered triangular waveforms sent to the envelope generator and the VCA.

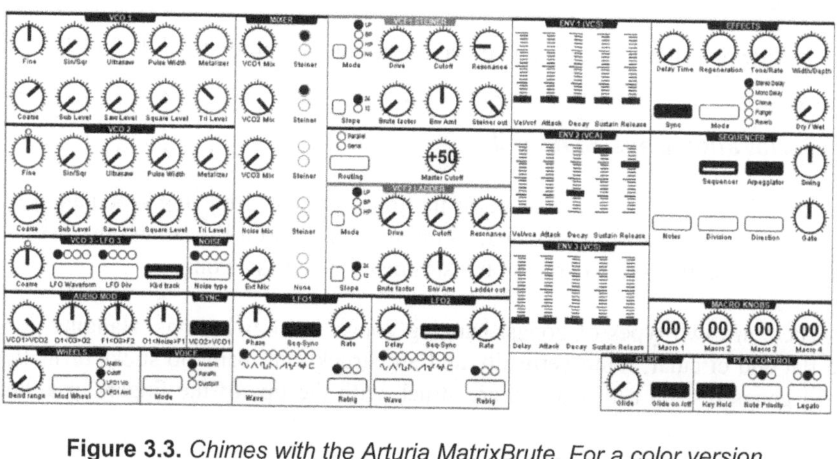

Figure 3.3. *Chimes with the Arturia MatrixBrute. For a color version of this figure, see www.iste.co.uk/reveillac/synthesizers2.zip*

Instructions and comments are as follows:

– The MatrixBrute does not explicitly have a ring modulator function, but we can quickly obtain it using the VCO 2 modulation control by VCO 1;

– turn the VCO 1 > VCO 2 control knob all the way to the right;

Figure 3.4. *The VCO 1 > VCO 2 button in the AUDIO MOD zone*

– set the appropriate frequencies for oscillators VCO 1 and VCO 2 as shown in Figure 3.3 (COARSE and TRI LEVEL);

– adjust the filter as shown (RESONANCE, ENV AMT and STEINER OUT); here, we opted for the STEINER filter (VCF 1), although the LADDER filter (VCF 2) would also have been suitable;

– adjust the VCA envelope as indicated (ENV 2);

– adjust the MASTER CUTOFF filter management control to obtain a value of +50;

– try pressing a key in the middle of the keyboard;

– the sound can be adjusted by adjusting the tuning of each of the VCOs (COARSE);

– a little reverb can be added to enhance the sound reproduction (EFFECTS).

3.1.3. *Novation Bass Station II*

Metallic sounds are the essence of the tones produced by ring modulation. For this new exercise, we will construct, using the Bass Station II, a sound that takes on its full scope in the Asian percussion instrument known as a gong.

There are two types of gongs: hanging gongs and bowl gongs. Hanging gongs are flat and circular, hung vertically and are struck with bamboo sticks or mallets. Bowl gongs rest on a tripod and are struck with the tips of the fingers or a small stick.

Gongs are generally made of bronze or brass, although other alloys can be used, which are ideal for the sound reproduction provided by a ring modulator.

The patch shown in Figure 3.5 simulates a small hanging gong, also called a ceremonial gong.

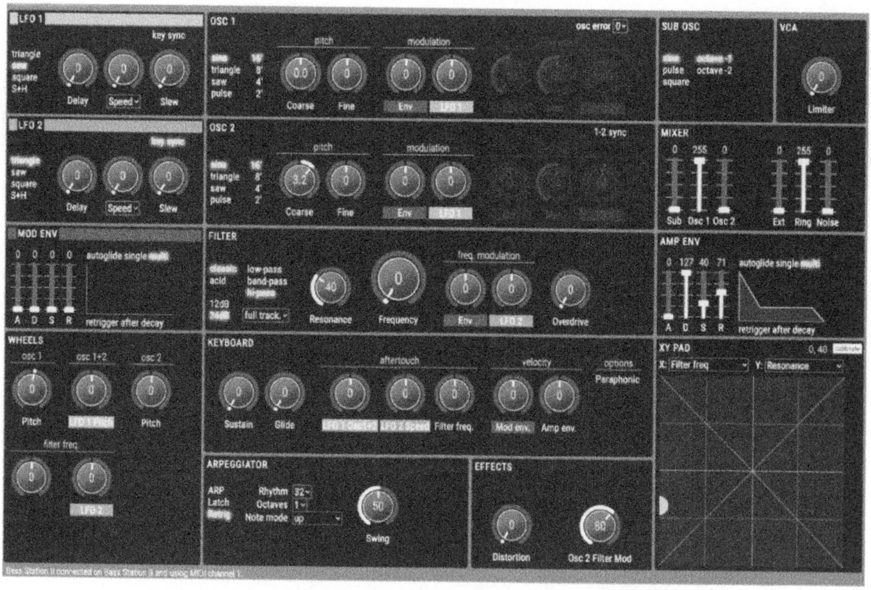

Figure 3.5. *A gong sound patch. For a color version of this figure, see www.iste.co.uk/reveillac/synthesizers2.zip*

Instructions and comments are as follows:

– set the values of the different controls as shown below;

– OSC 2: SINE 16', PITCH = 3.2;

– SUB OSC: SINE, OCTAVE-1;

– AMP ENV: A = 0, D = 127, S = 40, R = 71, MULTI ;

– FILTER: CLASSIC, HI-PASS, 24 dB, FULL TRACK;

– FILTER: RESONANCE = 40;

– MIXER: OSC 1 = 266, RING = 255;

– ARPEGGIATOR: default;

– all other parameters: 0;

– remember to set the output volume to an appropriate value.

The pitch setting of oscillator 2 and the filter's resonance can be adjusted according to the user's taste.

3.1.4. *Max/MSP*

Figure 3.6 shows a simple ring modulation application. A MIDI keyboard sends a note to an oscillator, which provides a sinusoidal signal (carrier). The latter is then modulated by a second oscillator, configurable in frequency and depth by the user.

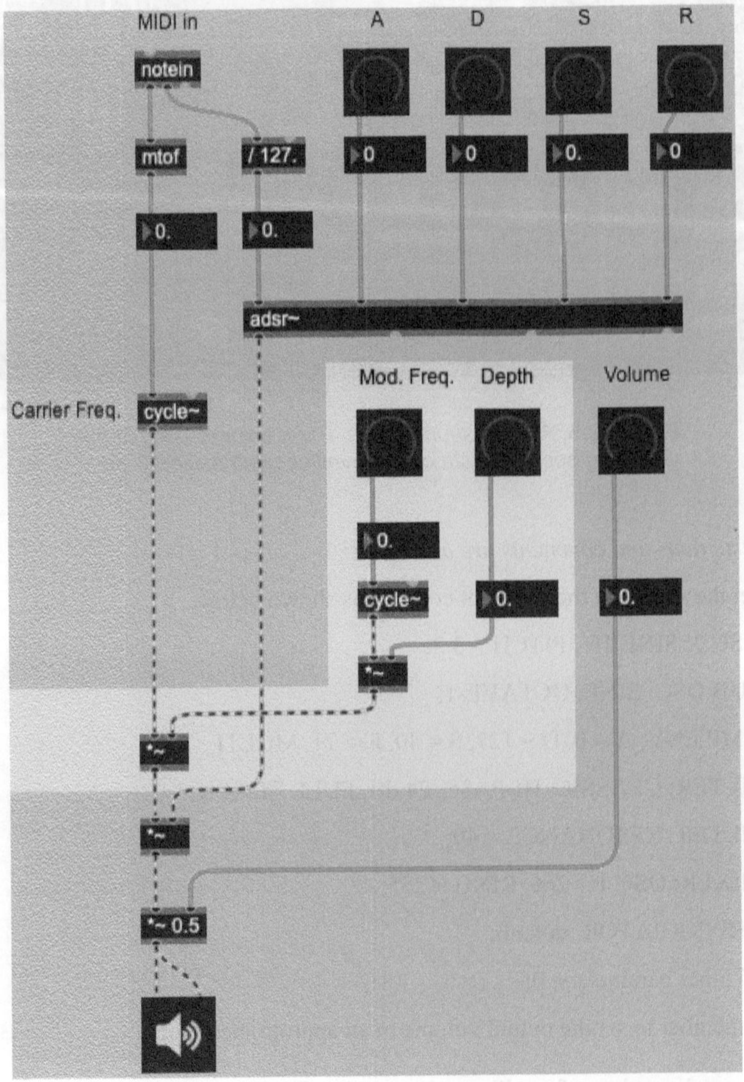

Figure 3.6. *Ring modulation*

An ADSR-type envelope is then applied to the modulated signal before joining the audio output (VCA).

Instructions and comments are as follows:

– the objects in the gray part of the patcher involve a classic VCO–ADSR setup (see section 1.4.1) and volume management (see section 1.4.3);

– the six remaining objects form the modulator;

– the `dial Mod. Freq.` object sets the modulating frequency. Its size is 1000, and its minimum value is 100;

– the `dial Depth` object sets the modulation depth. Its size is equal to 1, and its minimum decimal value is 0;

– the `number` object connected to the output of the `dial Mod. Freq.` object displays the value of the chosen frequency;

– the `cycle~` object whose input is connected to the output of the previous `number` object generates the sinusoidal signal of the modulator;

– the `number` placed behind the `dial Depth` object displays the depth value;

– the `*~` object combines oscillator signal and depth;

– the next `*~` object generates the product between the carrier signal and the modulating signal;

– do not forget to turn up the volume and set an envelope to produce a beep when you press a key on the keyboard.

The modulating frequency must be far from the carrier frequency to obtain the classic effect due to ring modulation, a sound reproduction like a chime.

By entering the following settings, you should get a decent sound: A = 50; D = 425; S = 0.675; R = 1999; Mod. Freq. = 475; depth = 0.8; and volume = 0.3. Do not hesitate to try other settings to get different sounds.

3.1.5. *Pure Data*

This exercise uses the patch presented in Figure 3.1. It has been improved so that a MIDI note can be chosen to define the carrier frequency. A modulating signal frequency between 1 and 1,000 Hz can also be chosen.

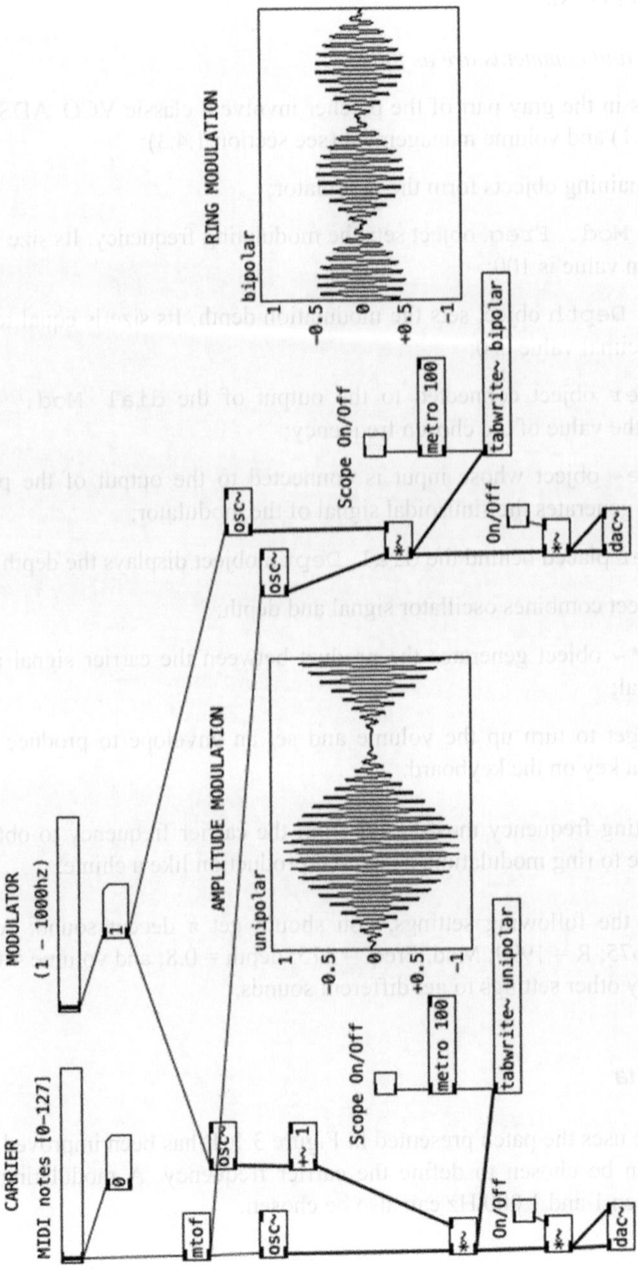

Figure 3.7. *Amplitude modulation and ring modulation*

Two oscilloscope screens display corresponding waveforms, so the user can compare amplitude modulation and ring modulation.

Instructions and comments are as follows:

– the construction of this patch does not reveal anything new compared to the exercises that we already covered in previous chapters;

– the left `slider` object allows you to choose a MIDI note between 0 and 127, which defines the carrier frequency;

– the left `slider` object allows you to choose a frequency according to a range of logarithmic values between 1 and 1,000, which defines the modulating frequency;

– we conventionally find the `number`, `mtof`, `osc~` objects to display and transform MIDI notes into frequencies;

– behind the `slider` object of the modulator, we also find a `number` object frequency display, followed by an `osc~` object to generate the modulating frequency;

– in the case of amplitude modulation, a `+~ 1` object makes the signal unipolar;

– the `*~` object combines the carrier signal and modulating signal;

– a `toggle` object, which is followed by a `*~` object, allows you to validate the passage of signals to the `dac~` objects that provide audio output;

– visualization is provided by two `array` objects with a size of 4,400, known as `unipolar` and `bipolar`, creating a Bezier curve with a scale of 0 and 4,399 on the x-axis and 2 and –2 on the y-axis;

– the two `toggle` objects ensure that the display is turned off or on for each oscilloscope;

– they are each followed by a `metro 100` object, which transmits the table fill values every 100 ms;

– the two `tabwrite~ unipolar`, and `tabwrite~ bipolar` objects, placed behind each `metro 100` object, ensure the writing of the signal curve in each of the windows defined by the `array` objects;

– modify the settings, listen to the sound reproduction and observe the modifications on both screens. You will see the differences between the waveforms that you can obtain.

3.1.6. *VCV Rack*

For this rack, we kept the three oscillator modules, the two envelope generator modules, the filter module, the mixer module and the VCA module. We added the ring modulation module, which is free from Bacon Music.

This module, called ALingADing, is very minimalistic. It has two inputs: one for the carrier signal (CAR) and one for the modulating signal (SIG). Only one output is available (OUT), and a button (MIX) modulates the chosen effects (DRY for little or no presence or WET for a significant presence).

The wiring is relatively standard compared to the exercises we have already covered.

Figure 3.8. *Modulated tubular chimes. For a color version of this figure, see www.iste.co.uk/reveillac/synthesizers2.zip*

Instructions and comments are as follows:

– take the rack created in the previous exercise (see Figure 2.34);

– remove a VCO module and its associated OCT module;

– add an ALingADing ring modulation module from Bacon Music;

– add an LFO module;

– connect the SIN output of the LFO module to the FM input of the first VCO, its frequency will modulate the signal;

– connect the TRI output of the first VCO to the GIG input of the ALingADing module to obtain the modulating signal;

– connect the SIN output of the second VCO to the CAR input of the ALingADing module to obtain the carrier signal;

– connect the OUT output of the ALingADing module to the IN1 input of the VCA MIX module;

– other connections remain unchanged;

– act on the frequency of the two VCOs to obtain the desired effect;

– remember to adjust the presence, DRY or WET, using the MIX button on the ALingADing module;

– increase the volume of input 1 on the VCA MIX module. This module is not essential; the output of the ALingADing module could be connected directly to the IN input of the VCF filter module;

– the LFO effect can be added by acting on the frequency adjustment button (FREQ) and the dosage of its effect with the FM button of the first VCO;

– as we have already specified, you can try the different outputs of the LFO and the two VCOs;

– there are other complex ring modulation modules in the VCV Rack library; do not hesitate to try them.

3.1.7. *Reaktor*

Installing a ring modulator in a Reaktor rack requires the installation of an additional module, which is not supplied with the Blocks Base library by default.

You must import the RING MODULATOR block available on the Web from the Reaktor User Library.

Other blocks providing ring modulation are also available, such as the AM106 Ring Modulator or Ringmod block. Do not hesitate to try them.

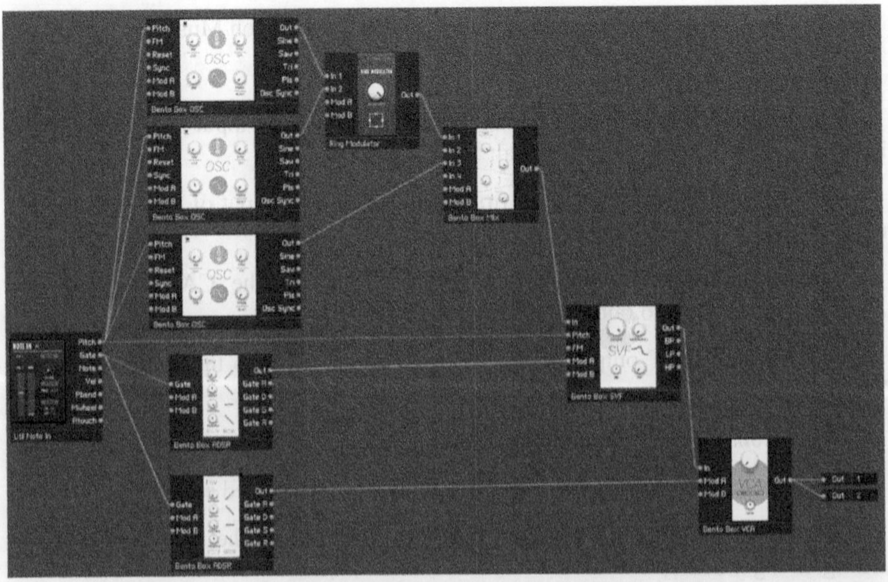

Figure 3.9. *Ring modulation with the Ring Modulator block. For a color version of this figure, see www.iste.co.uk/reveillac/synthesizers2.zip*

Instructions and comments are as follows:

– repeat the previous exercise (see section 2.3.6);

– remove the 2 LFO blocks and the mixing block placed in front of the VCA;

– add the RING MODULATOR block (MOD-Ring Modulator.ism);

– remove the connections between oscillators 1 and 2 and the remaining mixer block;

– connect the OUT output of the first oscillator to the IN 1 input of the RING MODULATOR block;

– connect the OUT output of the second oscillator to the IN 2 input of the RING MODULATOR block;

– connect the OUT output of the RING MODULATOR block to the IN 1 input of the mixing block;

– connect the OUT output of the SVF block to the IN input of the VCA block;

– detune the second oscillator relative to the first and turn the MOD TYPE knob on the RING MODULATOR block to the right to take the modulating signal into account;

– adjust the envelope generators to have a brief attack, a decay of at least 1 s, and a release close to 700 ms;

– the volume adjustment button 3 of the mixing block allows the sound signal coming from the ring modulator to be mixed with that of the third oscillator; you can vary its volume to your liking.

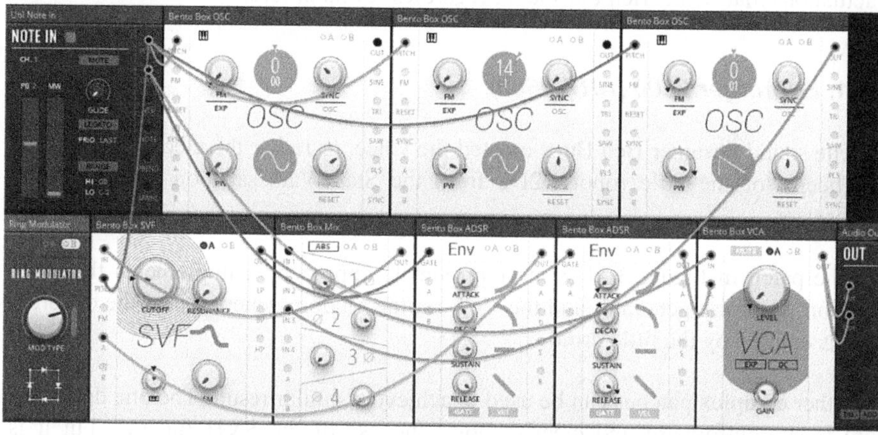

Figure 3.10. *Ring modulation in Panel mode. For a color version of this figure, see www.iste.co.uk/reveillac/synthesizers2.zip*

The settings in Figure 3.10 are shown for information purposes only. Do not hesitate to modify all the oscillator, filter and envelope settings to explore several sound rendering possibilities.

3.2. Exercise 9 – sample and hold

The sample and hold, S&H and S/H function is present on many synthesizers such as the ARP 2600, the ARP Odyssey, the Arturia MatrixBrute, the Roland System-100, the Moog 55 Modular, the Korg MS-20 and many others. Its purpose is to capture the amplitude of a signal at a given time and keep it constant over a certain period.

It is the basis of many realistic sounds and is one of the keys to humanizing digital audio signals, which are often cold and abstract. It is essential for imitating real instruments whose sound reproduction oscillates around the main note, such as specific brass instruments, percussion or strings. Finally, it is also the driving force behind numerous sound generators that evolve, respecting random criteria.

For all of these reasons, many LFOs integrated into synthesizers have a random waveform that can be used with S&H alongside classic shapes – sinusoidal, square, rectangular, triangular or sawtooth. The presence of a "digital chance" will add fluctuations that will enrich the sound, regardless of its timbre.

3.2.1. *Behringer 2600 (ARP 2600)*

The sound designer Ben Burtt created the voice, or rather the electronic whistles and beeps, of the little robot R2D2 from the Star Wars saga on an ARP 2600 synthesizer.

The patch in Figure 3.11 is an approximate approach; it implements the S&H function, the noise generator and the VCF for the random generation of the different signals emitted by the little droid.

Other complex patches can be used to achieve a similar result. The one described here is straightforward and clearly shows the use of the S&H function but it is perhaps not the most realistic in terms of sound reproduction.

Instructions and comments are as follows:

– using a patch cable, connect the S/H OUT output to the VCO2 (CONTROL) input of the VCF, as shown in Figure 3.11;

– set the S&H LEVEL to its maximum;

– set the filter resonance to its maximum;

– adjust VCA settings as shown;

– set the noise level (NOISE GENERATOR) to its maximum;

– set the mixer's VCA output level to your liking;

– by changing the color of the noise from pink to white (PINK–WHITE), you can slightly modify the timbre of the outgoing signal;

– play the S&H RATE and VCF INITIAL FILTER FREQUENCY controls to generate the appropriate tones.

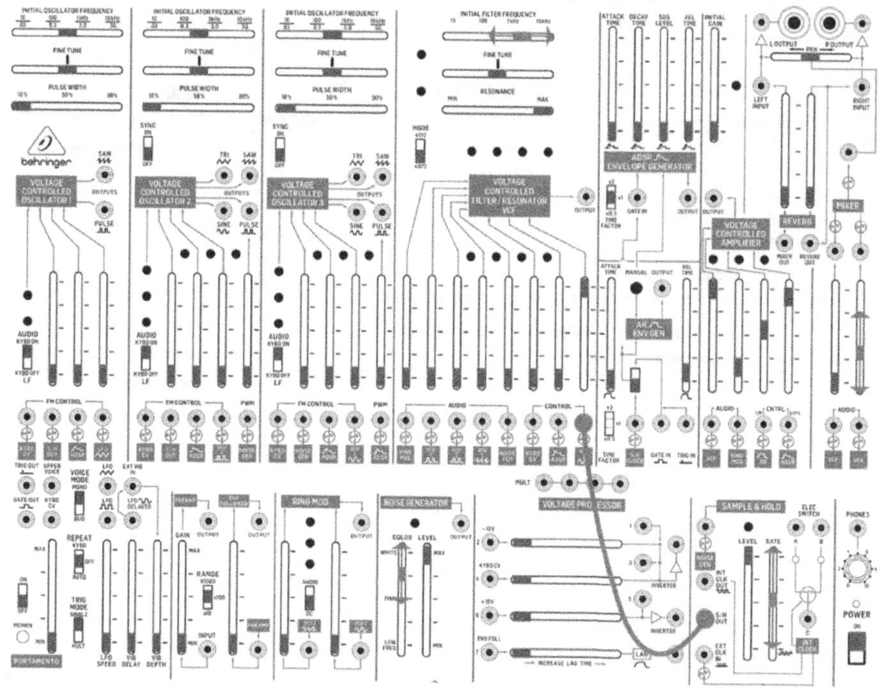

Figure 3.11. *An R2D2-style patch. For a color version of this figure, see www.iste.co.uk/reveillac/synthesizers2.zip*

3.2.2. *Arturia MatrixBrute*

The patch presented below, inspired by Diego Tejeida, was specially created for the MatrixBrute.

Tejeida was the keyboardist of the group Haken until 2021; he is also a sound designer and sound engineer and works from his studio, the Bunker, in Mexico.

To create this unique sound, the three oscillators and two filters of the MatrixBrute are used. LFO 1, placed in S&H mode, creates subtle variations in the gradual unfolding of the sound along with the different envelope generators when a key remains pressed. LFO 2, the noise generator, and specific effects are also present to magnify this patch's very complex sound rendering.

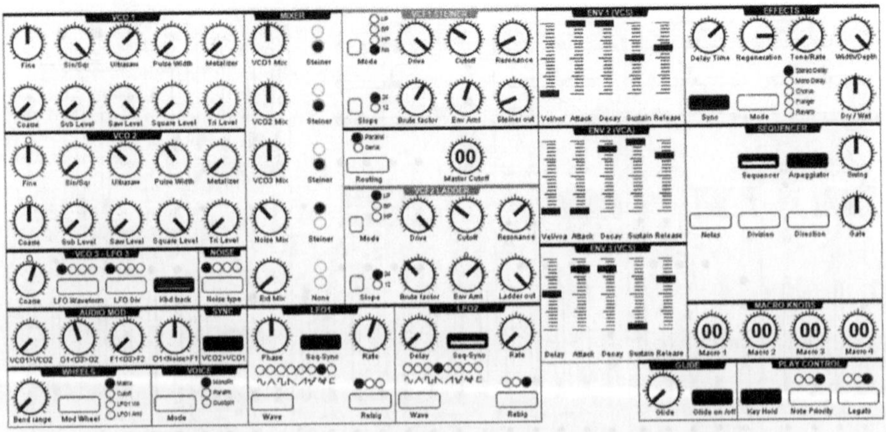

Figure 3.12. *A patch using S&H. For a color version of this figure, see www.iste.co.uk/reveillac/synthesizers2.zip*

Instructions and comments are as follows:

– associated with the settings in Figure 3.12, you will need to set up the assignments in Table 3.2 on the matrix;

From	To	From	To
ENV3	VCO1 Ultra	LFO2	VCO1 Ultra
ENV3	VCO2 PW	LFO2	VCO1 Metal
LFO1	VCO1 Pitch	LFO2	Steiner Cutoff
LFO1	VCO2 Pitch	LFO3	VCO1 Ultra

Table 3.2. *MatrixBrute Matrix Assignments*

– change the RATE setting of LFO 1 to see the influence of S&H;

– by altering the modulation level of the filters, you can modify the aggressiveness and roundness of the sound.

3.2.3. *Max/MSP*

For this exercise, we will capture the amplitude of a signal created by a noise generator whose amplitude varies between –1 and 1 at a given moment. The sampling is triggered regularly by a sawtooth oscillator whose frequency we can vary.

The imposed trigger threshold is set at 0.5.

A note generated by a MIDI keyboard is transformed into a frequency, which is then combined with the sampled value to vary the oscillator frequency, providing a sinusoidal waveform.

The user can adjust the overall volume.

Instructions and comments are as follows:

– reproduce the patcher in Figure 3.13;

– the `noise~` object generates the noise signal;

– the three `dial`, `number` and `phasor~` objects control the trigger sawtooth signal;

– the `dial Rate` object has a size of 1 and a minimum value of 21;

– the `sah~` object performs the S&H function with an imposed threshold of 0.5;

– the `+~ 1` object makes the signal unipolar (>1) so it can process it correctly;

– the `*~` object multiplies the signal coming from the MIDI keyboard with the sampled value;

– the `saw~` object generates the sawtooth signal directed toward the `slider gain` object, which controls the audio output volume;

– the three `notein`, `mtof` and `flonum` objects capture the note pressed on the keyboard, transform it into a frequency and then display it;

– the three `number~` objects display the sampled value (positive or negative), the frequency multiplier (>0) and the resulting frequency, which will be provided by the sawtooth oscillator (`saw~`);

– Figure 3.14 uses the patcher in Figure 3.13 and shows, using several oscilloscopes, the evolution of the different signals at time t;

– adjust the frequency and volume and choose a note on your keyboard to listen to the S&H function working.

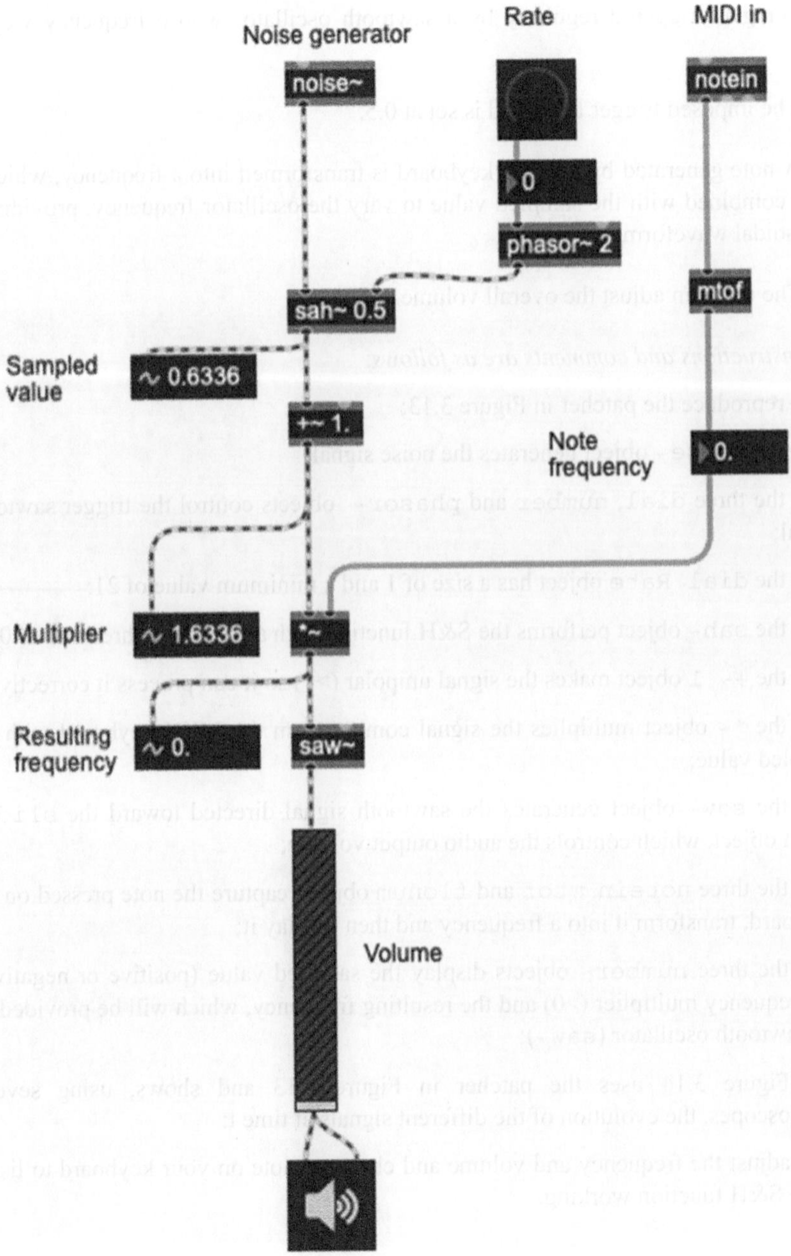

Figure 3.13. *Sample and hold with Mx/MSP*

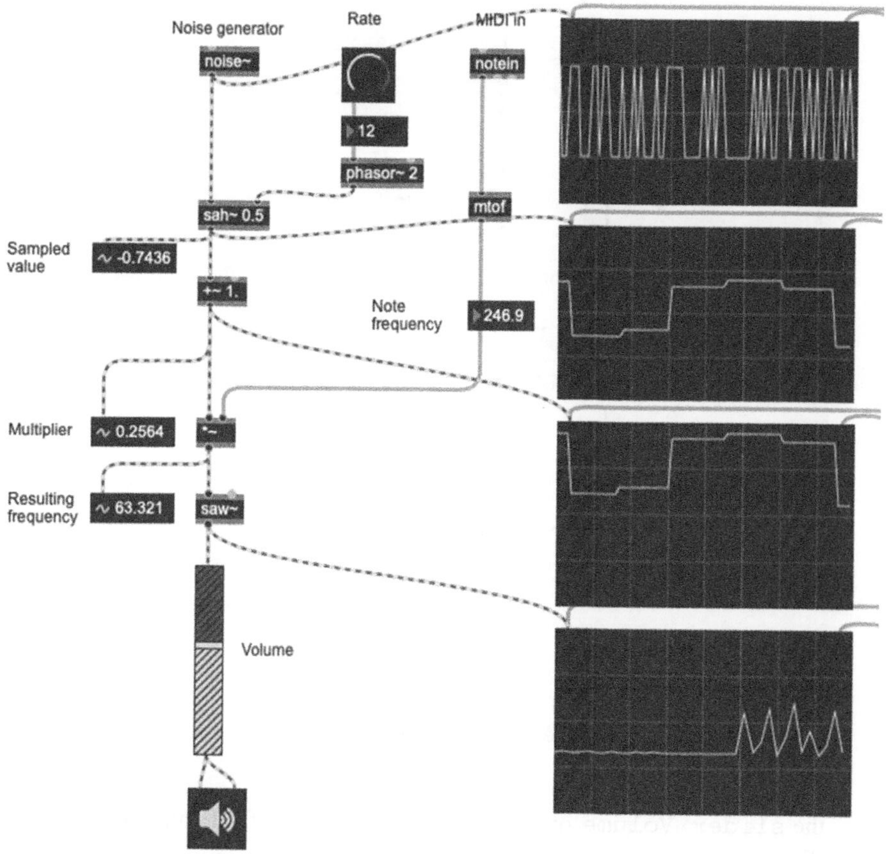

Figure 3.14. *The S&H patcher and its different signals*

3.2.4. *Pure Data*

For this exercise, the Pure Data patch is straightforward; it uses the classic configuration of an S&H function, a noise generator within which the series of values of different amplitudes will be sampled following a frequency determined by an oscillator generating a sawtooth signal. You can control the overall output volume.

Instructions and comments are as follows:

– create the patch in Figure 3.15;

– the `slider Rate` object handles a range of linear values between 0 and 50;

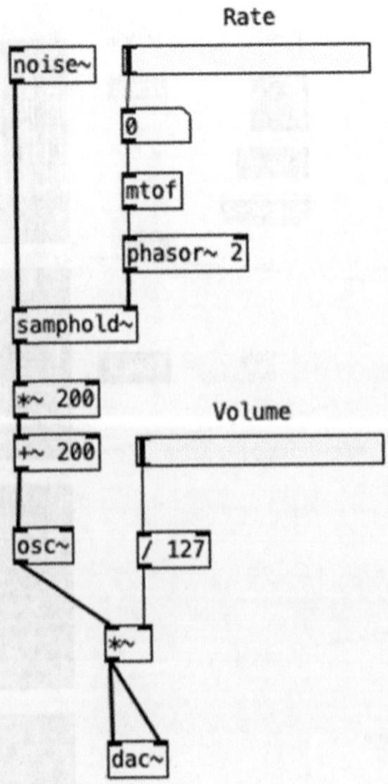

Figure 3.15. *S&H with Pure Data*

– the `slider Volume` object manages a range of default values (linear from 0 to 127);

– the `noise~` object manages a range of default values (linear from 0 to 127);

– the `mtof` object converts the value generated by the `slider Rate` object in a frequency usable by the `phasor~` object that follows;

– the `samphold~` object executes sampling at a speed determined by the `slider Rate` object via a sawtooth signal created by the `phasor~ 2` object;

– the `*~ 200` object multiplies the `samphold~` object output signal by 200 to obtain an audible frequency;

– the `+~ 200` object makes the signal unipolar to be used correctly by the oscillator `osc~` object that follows;

– the other objects follow the typical sequence already encountered in the previous exercises;

– do not forget to set the `Volume` to an appropriate value;

– you can change the `Rate` to vary the sampling rate;

– by changing the value 200 of the object `*~` (for example, 880) you will get a different average pitch.

3.2.5. *VCV Rack*

We downloaded the S&H module from Count Modula, developed by Adam Verspaget under license GNU GPLv3 for this rack.

An LFO with a sawtooth signal triggers the S&H module trigger, which samples a random amplitude value captured within a white noise signal. The S&H module then controls a VCO module, which generates a triangular signal directed to a VCF and then amplified by a VCA. Two envelope generators – for the VCA and VCF – condition the temporal shapes of the output signal when a key is pressed on the keyboard.

Figure 3.16. *A noise, LFO and S&H rack. For a color version of this figure, see www.iste.co.uk/reveillac/synthesizers2.zip*

Instructions and comments are as follows:

– insert the following modules into the rack: MIDI-CV, LFO, NOISE, VCO, ACV, AUDIO and VCF;

– add two ADSR modules and the Count Modula S&H module that you previously downloaded;

– connect the inputs–outputs of the different modules, as shown in Figure 3.16 and Table 3.3. The envelope generators are differentiated according to their destination, VCF or VCA, on the left and right, respectively, in Figure 3.16;

From	To	From	To
MIDI-CV–GATE	ADSR (VCF)–GATE	**NOIS–WHIT**	S&H–INPUT
MIDI-CV–GATE	ADSR (VCA)–GATE	**NOIS–OUTPUT**	VCO–V/OCT
LFO–SAW	S&H–TRIGGER	**VCO–TRI**	VCF–IN
ADSR (VCF)–ENV	VCF–CUT	**VCF–LPF**	VCA–IN
ADSR (VCA)–ENV	VCA–Control	**VCA–OUT**	AUDIO–L/MON

Table 3.3. *Connections between the different modules for exercise 9*

– do not forget to adjust the VCA volume and output level;

– the frequency of the LFO controls the sampling speed of the S&H module;

– it is better to set a long release (REL) on the envelope generators for a continuous note after you release the keyboard key;

– choose mode S of the S&H module for no white noise in the output signal. T mode is for Track & Hold mode, and P mode is for Pass & Hold mode.

3.2.6. *Reaktor*

Rather than reproducing the standard configuration, noise generator and LFO, the rack in Figure 3.17 unusually uses an S&H module.

Here, the sampled source signal and the trigger signal are generated by an LFO, allowing us to generate a repetitive melody without it being completely random.

A control oscilloscope allows the user to view the output signal from the S&H module to understand the influence of the S&H function on the output signal provided by a VCA preceded by a VCO, following the classic setup.

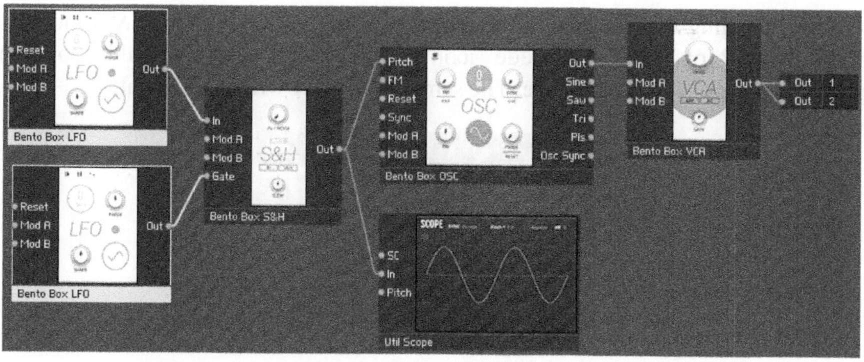

Figure 3.17. *A rack using S&H. For a color version of this figure, see www.iste.co.uk/reveillac/synthesizers2.zip*

Instructions and comments are as follows:

– create a new rack;

– insert two MOD-BentoLFO modules, one PRO-Bento S&H module, one OSC-Bento Oscillator module, one AMP-Bento VCA module and one UTIL-Util Scope module;

– connect the OUT output of the first LFO module to the GATE input of the S&H module to accommodate the signal that will be sampled;

– connect the OUT output of the second LFO module to the IN input of the S&H module, which will receive the sampling trigger signal;

– connect the OUT output of the S&H module to the PITCH input of the VCO module and to the IN input of the SCOPE module to be able to visualize the shape of the output signal;

– connect the OUT output of the VCO module to the IN input of the VCA module on which you can adjust the volume and gain;

– connect the OUT output of the VCA to the two inputs of the AUDIO OUT output module;

– set the PITCH of the SCOPE module to 64, the sync mode to SLOPE and the gain in dB to 3;

– choose a square waveform on the first LFO (connected to the GATE of the S&H module);

– by altering the two frequency settings of the LFO modules, you will be able to construct a sound melody;

– by switching the second LFO (connected to the IN input of the S&H module) to UNIPOLAR or BIPOLAR mode (icon in the top-left of the module), you can modify the output sound sequence. The sampling trigger will be different;

Figure 3.18. *The bipolar (left) or unipolar (right) toggle icon (of the LFO module. For a color version of this figure, see www.iste.co.uk/reveillac/synthesizers2.zip*

– by modifying the waveform of this same LFO (rising or descending sawtooth, sine, triangle, etc.), you will vary the output sound sequence;

– you can modify the selection of the VCO module note and its waveform to modify the general tone of the sound sequence. Remember to check that the VCO module is in Key Tracking mode (keyboard icon visible in the top-left of the module, see Figure 3.19);

Figure 3.19. *In the top-left, to the right of PITCH, the Key Tracking icon of the VCO module. For a color version of this figure, see www.iste.co.uk/reveillac/synthesizers2.zip*

– do not forget to adjust the volume and gain on the VCA module;

– try out the settings in Table 3.4; you will obtain some unique melodic sequences. Many other sequences can be tried out by adjusting the different LFO and VCO settings;

– the S&H module has its own internal white noise generator, which can be activated by turning the IN/NOISE knob. When activated, a completely random signal is superimposed onto the existing signal and the output sound melody becomes random.

LFO 1 (GATE input of S&H)	LFO 2 (IN input of S&H)	VCO
Bipolar – Freq: 8 – Square	Bipolar – Freq: 18 – Descending sawtooth	Note: 0 – Triangular
Bipolar – Freq: 12 – Square	Bipolar – Freq: 18 – Descending sawtooth	Note: 24 – Sinus
Bipolar – Freq: 10 – Square	Unipolar – Freq: 24 – Triangular	Note: 0 – Triangular
Bipolar – Freq: 16 – Square	Unipolar – Freq: 54 - Triangular	Note: 0 – Triangular
Bipolar – Freq: 15 – Square	Unipolar – Freq: 100 – Descending sawtooth	Note: 0 – Triangular
Bipolar – Freq: 15 – Square	Bipolar – Freq: 45 – Descending sawtooth	Note: 0 – Triangular

Table 3.4. *Possible rack configurations*

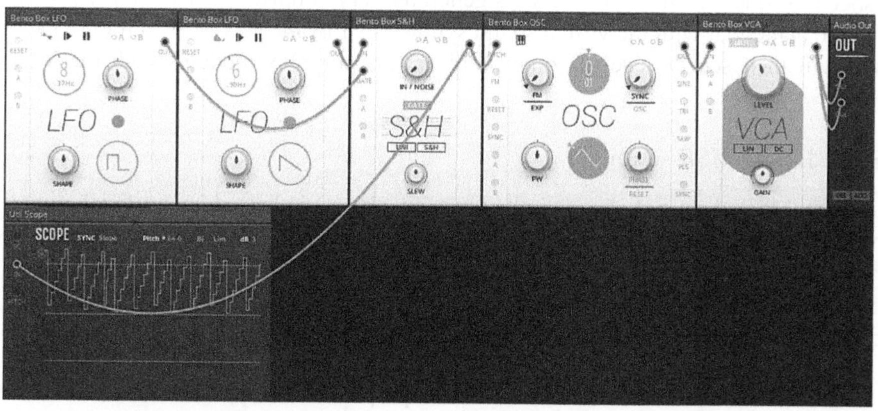

Figure 3.20. *The S&H rack in Panel mode. For a color version of this figure, see www.iste.co.uk/reveillac/synthesizers2.zip*

3.3. Sound effects

Sound effects have become indispensable over time. Users even go so far as to use effects pedals intended for guitarists by connecting them to the audio output of their synthesizers.

Manufacturers are not left out; many hardware or virtual machines have sound effects. Reverberation is present on many current and vintage synthesizers and is the most common effect.

Distortion, flanger, phaser, chorus and delay can also be present, but other less conventional effects like the pitch shifter or the octaver can also exist on modular machines.

Sections 3.3.1–3.3.4 will discuss these, although some of them, as mentioned previously, can be added as pedals or multi-effect racks at the output of your instrument.

3.3.1. *Exercise 10 – reverberation*

The reverberation (reverb) discussed in section 3.8.4 of Volume 1 is available on many synthesizers such as the ARP 2600, the Arturia MatrixBrute, etc. As for modular synthesizers, there are many modules dedicated to this purpose.

Presenting patches using reverb on hardware synthesizers is not of much interest, especially since its implementation is generally very simple. We will simply present how to achieve it with the virtual sound synthesis software Max/MSP and Pure Data. We will then see the use of some more or less complex modules with the VCV Rack.

3.3.1.1. *Max/MSP – Reverb 1*

Max/MSP does not, by default, have an object dedicated to artificial digital reverb, so it must be reconstructed by applying the theoretical bases on which it is relies. There are several ways to create this effect, the best known and most used in most commercial systems is based on the work of Manfred Robert Schroeder[1], dating back to 1960–1970.

To create artificial reverb, Schroeder proposed an approach based on the combination of comb filters and all-pass filters. One of the main problems encountered with this approach is the metallic timbre the sound signal takes on.

To solve this problem while maintaining an acceptable reverb density, Schroeder proposed using two combinations: all-pass filters in series and comb filters in parallel. This solution is imperfect because its frequency response is not linear but is sufficient to obtain the correct reverb.

The patcher in Figure 3.21 works with a “SchroederVerb” abstraction, which contains all-pass filters and comb filters; for the rest, you will find sequences of objects seen previously.

---

1 Manfred Robert Schroeder (1926–2009) was a German physicist known for his work on acoustics and computer graphics.

A MIDI control section directs a sawtooth signal, corresponding to the note chosen on the keyboard, to a low-pass filter whose cutoff frequency and resonance are adjustable by the user, and the audio output can be adjusted using the master volume control.

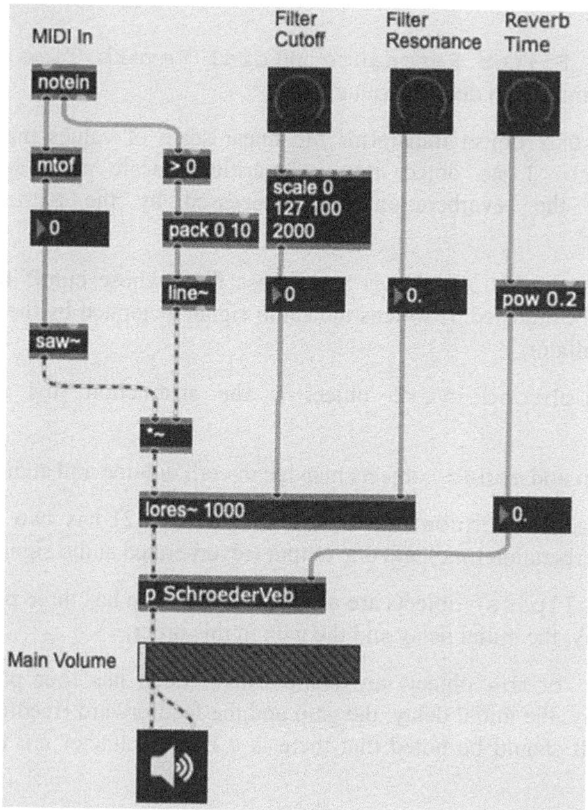

Figure 3.21. *A Max/MSP patcher to control the reverb*

Instructions and comments are as follows:

– the objects on the left part of the patcher, `notein`, `mtof`, `number`, `saw~`, `> 0`, `pack 0 10`, `line~` and `*~` capture the note coming from the MIDI keyboard, transform it into a frequency that controls the sawtooth oscillator while taking into account the keyboard velocity to limit the note length. This procedure has already been encountered (see section 2.2.3, exercise 4b);

– the right part does not have any new items either; you will find three dial objects, one value scale recalculation scale object, one object pow of compute the hth power and three value control flonum objects;

– the dial Filter Cutoff object controls a range of values from 0 to 127 (default);

– the dial Filter Resonance and dial Reverb Time objects have a size of 0 and a minimum decimal value of 1;

– the pow 0.2 object transforms the linear scale of values managed by the dial Reverb Time object into a logarithmic scale more adapted to the adjustment of the reverberation time processed by the SchroederVerb abstraction;

– the lores~ 1000 object is a low-pass filter whose cutoff frequency and resonance can be adjusted. It softens the audio signal generated by the saw~ object, a sawtooth oscillator;

– the p SchroederVerb object is the abstraction that generates the reverberation;

– the gain and ezdac~ objects manage overall volume and audio output;

– the abstraction SchroederVerb (see Figure 3.22) has two inputs (audio signal and reverberation time) and one output (reverberated audio signal);

– the four allpass~ objects are all-pass filters. Each has three parameters: the maximum delay, the initial delay and the gain in this order;

– the eight comb~ objects are comb filters. Each has four parameters: the maximum delay, the initial delay, the gain and the feedforward (feedforward signal) in this order. It should be noted that there is a fifth parameter not used here: the feedback;

– the *~ 0.1 object reduces the amplitude of the incoming audio signal by 10;

– the parameters of the different filters were calculated as accurately as possible following the Schroeder method (see the reference section at the end of the book). You will find them brought together in Tables 3.5 and 3.6;

– do not forget to turn up the general volume, otherwise no sound signal will be heard;

– adjust the different settings of the low-pass filter and the reverberation time to listen to their influence on the sound reproduction.

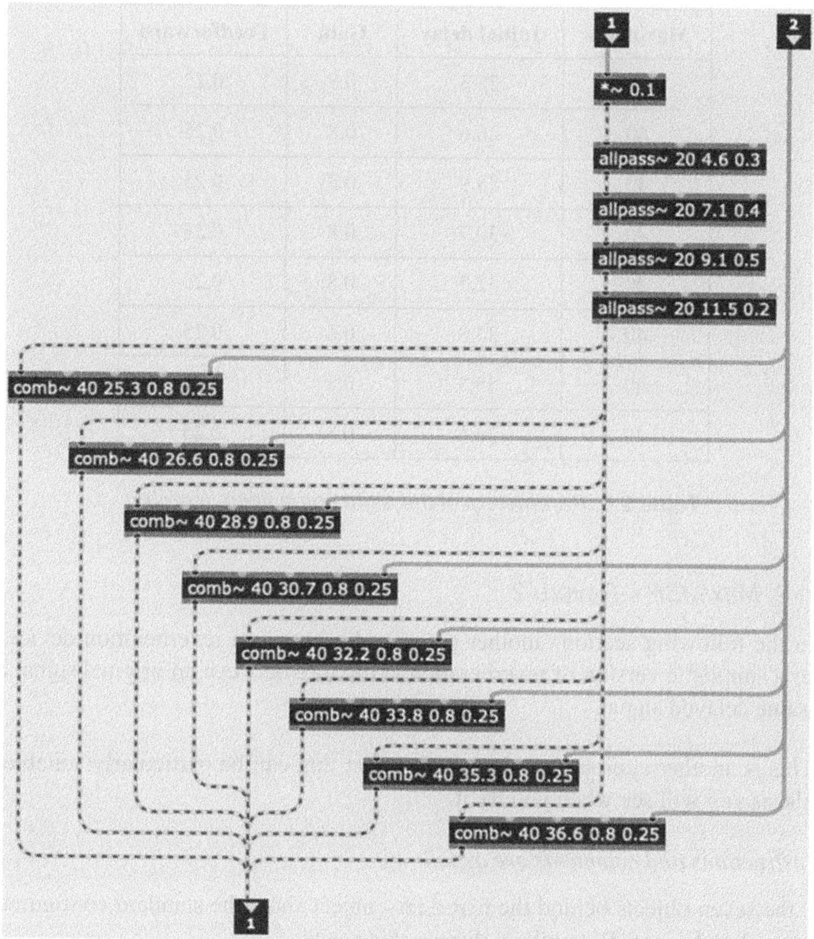

Figure 3.22. *The "SchroederVerb" abstraction used by the patcher*

Maximum delay	Initial delay	Gain
20	4.6	0.3
20	7.1	0.4
20	9.1	0:5
20	11.5	0.2

Table 3.5. *Parameter of the four all-pass filters (allpass~)*

Max delay	Initial delay	Gain	Feedforward
40	25.3	0.8	0.25
40	26.6	0.8	0.25
40	28.9	0.8	0.25
40	30.7	0.8	0.25
40	32.2	0.8	0.25
40	33.8	0.8	0.25
40	35.3	0.8	0.25
40	36.6	0.8	0.25

Table 3.6. *Parameters of the eight comb filters (comb~)*

3.3.1.2. *Max/MSP – Reverb 2*

In the following section, another patcher also acts as a reverberation device, or rather a simplistic version of reverberation, a mixture between an original signal and this same delayed signal.

This is another approach to the reverb effect that can be particularly suitable on vocals, as you will see when you try it.

Instructions and comments are as follows:

– the seven objects behind the `notein~` object show the standard configuration we have already seen. We will not discuss this again;

– to the left, the umenu, which has three options: `None`, `Micro` and `Keyboard`, allows the user to choose the type of incoming signal, either a voice captured by a microphone or a note from a MIDI keyboard;

– the `ezacd~` object initiates sound capture from an external source, usually the microcomputer microphone. This object is of the on/off type. A user click can activate it;

– the `selector~` object validates the user input (`None`, `Micro` or `Keyboard`) by directing the chosen input toward its output;

– the right part is the part for the reverb or rather delay;

– the four `slider` objects allow the user to alter the various settings;

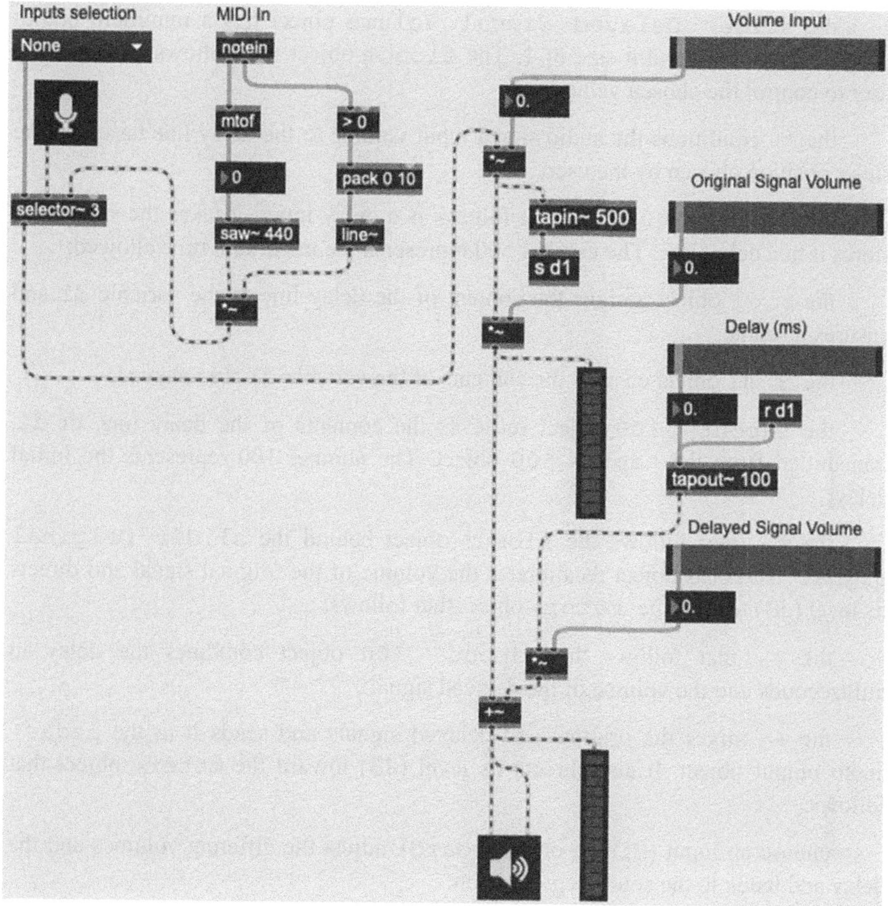

Figure 3.23. *A simulated reverb with a delay*

– the `slider Volume Input` object has a minimum output decimal value of 0 and a size of 1;

– the `flonum` object that follows it allows the user to control the chosen value;

– the `slider Original Signal Volume` object has a minimum output decimal value of 0 and size of 1. The `flonum` object that follows allows the user to control the chosen value;

– the `slider Delay (ms)` object has a minimum output value of 0 and a size of 501. The `flonum` object that follows it allows the user to control the chosen value;

– the slider Delayed Signal Volume object has a minimum output decimal value of 0 and a size of 1. The flonum object that follows it allows the user to control the chosen value;

– the *~ conditions the audio signal input volume to the delay line based on the slider position chosen by the user;

– the tapin~ 500 object that follows is a delay input; it takes the signal and stores it in a delay line. The number 500 represents the maximum time allowed;

– the s d1 object assigns the content of the delay line to the variable d1 and ensures it sends;

– the r d1 object ensures the contents of the variable d1 are received;

– the tapout~ 100 object retrieves the contents of the delay line, or d1, transmitted from the tapin~ 500 object. The number 100 represents the initial delay;

– the *~ that follows the flonum object behind the slider Original Signal Volume object recalibrates the volume of the original signal and directs its level (dB) toward the meter~ object that follows;

– the *~ that follows the tapout~ 100 object combines the delay in milliseconds and the volume of the delayed signal;

– the +~ mixes the original and delayed signals and sends it to the ezdac~ audio output object. It also directs its level (dB) toward the meter~ object that follows;

– choose an input (Micro or Keyboard), adjust the different volumes and the delay and listen to the sound reproduction;

– if the fader of the slider Original Signal Volume object is at zero, you only hear the delayed signal.

NOTE.– If the duration of the delay is long (>100 ms), the reverb, so to speak, transforms into an echo, repeating only once.

3.3.1.3. *Pure Data – Reverb 1*

For this exercise, we will use the Pure Data[2] external rev2~ function, which is present by default in version 0.52-1 (Miller Puckette's "vanilla" distribution) used in this book.

---

2 Available at: www.puredata.info/docs/ListOfPdExternals/.

Older versions also implemented this feature. To check if it is present, go to the Help menu in Pure Data, then the List of objects submenu, etc., and check, in the EXTRA section, that the reverberator objects, `rev1~`, `rev2~` and `rev3~` are present (see Figure 3.24).

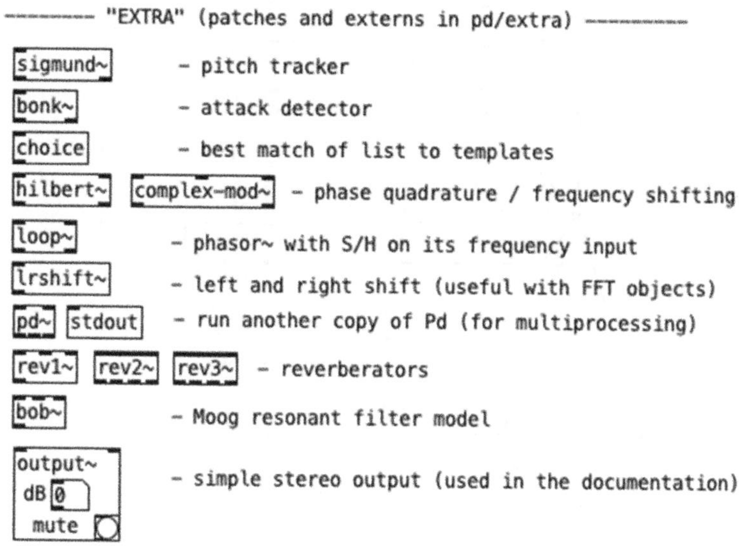

Figure 3.24. *List of objects present in the EXTRA section*

The patch corresponding to the first part of this ninth Pure Data exercise is presented in Figure 3.25. It allows you to test one of the reverb functions via a note sent from a MIDI keyboard.

The user can adjust the reverb level, internal feedback percentage, cutoff frequency and HF damping, the latter two working together.

When the processed frequencies are higher than the cutoff frequency, the feedback is reduced by the damping percentage. If the latter is zero, the reverberation time is equal for all frequencies; however, at 100%, almost no signal remains above the cutoff frequency.

Two other settings are available: a dry/wet balance, which balances the presence of the reverb effect in relation to the original signal (dry without effect, wet with effect), and a main volume.

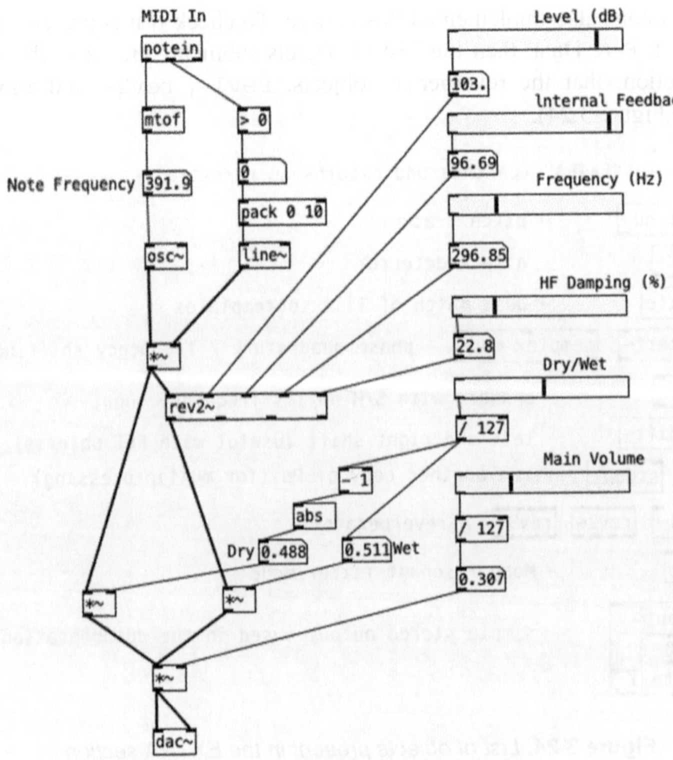

Figure 3.25. *Reverb with Pure Data*

Instructions and comments are as follows:

– the `notein`, `mtof`, `> 0`, `number`, `pack 0 10`, `osc~`, `line~` and `*~` objects located at the top left of the patch in Figure 3.25 have already been addressed in section 2.2.4 and will not be discussed further. They transmit a note in the form of a sinusoidal audio signal when a user presses a note on the MIDI keyboard;

– the objects located on the right manage all user settings available. Each `slider` object is connected to a `number` object to check the chosen value;

– the `slider Level (dB)` object handles a range of linear values between 0 and 120;

– the `slider Internal Feedback` object manages a range of linear values between 40 and 100. If the value is equal to 100, the reverb becomes infinite; at 90 the reverb is long and at 80 it becomes very short;

– the `slider Frequency (Hz)` object manages a range of linear values between 50 and 1000;

– the `slider HF Damping (%)` object manages a range of linear values between 0 and 100;

– the `slider Dry/Wet` object handles a range of linear values between 0 and 127;

– the `slider Main Volume` object manages a range of linear values between 0 and 127;

– the `- 1` and `abs` objects make it possible to balance the value transmitted by the `slider Dry/Wet` object in order to manage the balance between the original signal and the reverberated signal, as shown with the `Dry` and `Wet number` objects. The `abs` object gives the absolute value of a number. Note that, to ensure this balance, other solutions are possible (see section 3.3.4.3);

– the `*~` objects ensure the mixing of the original signal and reverberated signal;

– the remaining ones provide audio output and overall volume control;

– try the different settings. Choose a correct reverb level to have a convincing effect and look at the influence of the cutoff frequency and damping depending on the frequency of the note generated by the `osc~` object behind the `mtof` object.

3.3.1.4. *Pure Data – Reverb 2*

For this second part of exercise 9 devoted to reverb, we will again use an external function attached to the object `freeverb~`.

This function needs to be installed to work. To install it, go to the Pure Data Help menu, choose Find Externals, type `freeverb~` in the search box, and click the Search button; when `freeverb~` appears, click on the package to install it.

The `freeverb~` object is quite sophisticated and has eight parameters (see `freeverb~` help); in the patch in Figure 3.26, we use six of them that the user can adjust as they wish:

– the bypass parameter allows you to bypass the reverb effect and therefore keep an unmodified original audio signal;

– the freeze parameter captures the end of the reverb when the audio signal is maintained, generating a sort of looping effect;

– the wet and dry parameters validate or invalidate the presence of the reverberation;

– the room size parameter determines the reverb time (dimensions of the reverberating room);

– damping sets the reverb damping.

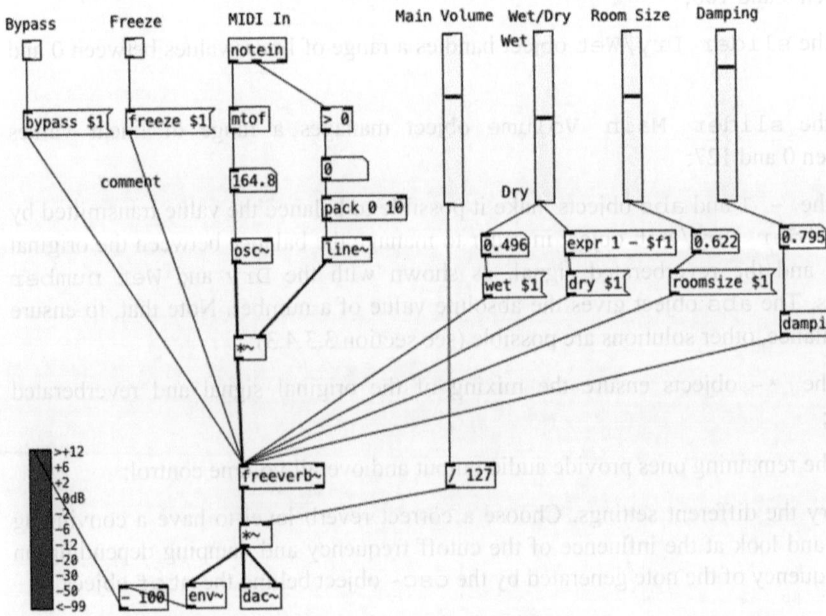

Figure 3.26. *A Pure Data patch using the* `freeverb~` *external function*

Instructions and comments are as follows:

– as in the previous patch, the `notein`, `mtof`, `> 0`, `number`, `pack 0 10`, `osc~`, `line~` and `*~` objects located in the central part of the patch in Figure 3.26 bring the audio signal coming from the MIDI keyboard;

– the `toggle` object and the `bypass $1` message activate or deactivate the bypass;

– the `toggle` object and `freeze $1` message enable or disable reverb freeze;

– the `freeverb~` object calls the external reverb function;

– the `slider Main Volume` objects and the `/ 127` object manage the main volume;

– the three `slider` objects (`Wet/Dry`, `Room Size`, and `Damping`) manage the reverberation settings;

– the three previous `slider` objects manage ranges of linear values between 0 and 1;

– the `expr 1 - $f1` object calculates the Dry mode value transmitted to the `message dry $1` object then the `freeverb~` object;

– the message `wet $1` object transmits the Wet mode value to the `freeverb~` object;

– the remaining objects have already been seen and provide audio output as well as calculation and display of the output level (see section 2.4.5);

– choose a correct volume level, do not forget to uncheck the bypass to hear the reverb effect and try the different settings.

3.3.1.5. *VCV Rack*

For this new rack, we will test several reverberation modules to compare their efficiency and settings, but above all, their sound coloring, which is often very different from one module to another.

We find a standard configuration for these reverberations: a VCO, a VCF, two envelope generators and a VCA.

We will add a muting module to disable the passage of the audio signal through the reverbs and a mixing module to manage levels and make different mixes.

The four reverberation modules chosen are as follows:

– the AS–Reverb FX module, developed by Alfredo Santamaria. Three adjustable settings were used: Decay, Damp and Blend, and + a Bypass function–monophonic (MIT license);

– the Chowdsp–ChowFDN module, developed by Jatin Chowhury. Six adjustable parameters were used: Pre-delay, Size, T60 Low, T60 High, Delays, Dry/Wet–monophonic (GPL 3.0 or + license);

– the RCM–GVerb module, developed by RCM. Eight adjustable parameters were used: Room Size, Time, Early, Damp, Bandwidth, Late, Mix, Spread + a Reset function–stereophonic (GPL 2.0 or + license);

– the Bidoo –REI module, developed by Bidoo. Seven adjustable parameters were used: Size, Damp, Width, Feedback, Pitch, Dry, Wet + a Freeze function and a Clipping function–stereophonic (GPL 3.0 or + license).

NOTE.– To study their characteristics in detail and have access to their user manuals, consult the publisher/author website from the VCV Rack library.

There are many other sophisticated modules. The ones above were selected for their performance, and they are free. After downloading and adding them to this rack, you can try the others at your leisure.

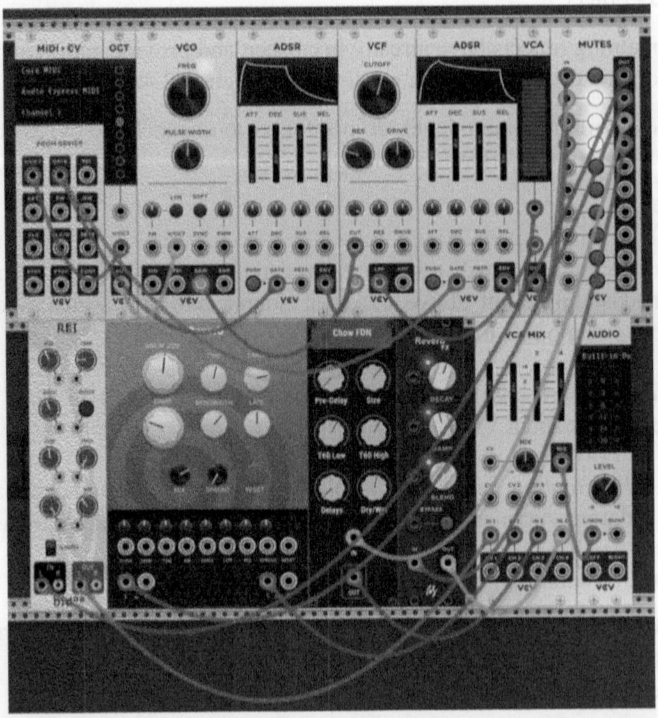

Figure 3.27. *Four reverb modules in one rack. For a color version of this figure, see www.iste.co.uk/reveillac/synthesizers2.zip*

Instructions and comments are as follows:

– create a new rack and add the following modules: MIDI-CV, OCT, VCO, 2 × ADSR, VCF, MUTES, VCA MIX, AUDIO and the four reverb modules;

– connect the inputs–outputs of the different modules as shown in Figure 3.27 and Table 3.7. The envelope generators are differentiated according to their destination, VCF or VCA, respectively, on the left and right in Figure 3.27;

– do not forget to adjust the VCA volume and output level on the AUDIO module;

– the buttons on the MUTES module allow you to open (gray) or close (white) the signal passage to one of the reverb modules;

– the sliders of the VCA MIX module allow you to adjust the volume of the signal coming from each reverberation module. They can also mix multiple signals if the passages have been opened from the MUTES module;

– do not hesitate to test the different reverberation modules with equivalent parameters to hear the different sound renderings;

– listen to the corrections made by the different settings of each reverb.

From	To	From	To
MIDI-CV–V/OCT	OCT–V/OCT	**VCA–OUT**	MUTES–IN 4
MIDI-CV–GATE	ADSR (VCF)–GATE	**MUTES–OUT 1**	REI–IN L
MIDI-CV–GATE	ADSR (VCA)–GATE	**MUTES–OUT 2**	GVERB–IN L
OCT–OUT	VCO–V/OCT	**MUTES–OUT 3**	CHOW FDN–IN
VCO–SAW	VCF–IN	**MUTES–OUT 4**	REVERB FX - IN
ADSR (VCF)–ENV	VCF–CUT	**REI–OUT L**	VCA MIX–IN 1
VCF–LPF	VCA–IN	**GVERB–OUT L**	VCA MIX–IN 2
ADSR (VCA)–ENV	VCA–CONTROL	**CHOW FDN–OUT**	VCA MIX–IN 3
VCA–OUT	MUTES–IN 1	**REVERB FX–OUT**	VCA MIX–IN 4
VCA–OUT	MUTES–IN 2	**VCA MIX–MIX**	AUDIO–L/MON
VCA–OUT	MUTES–IN 3	–	–

Table 3.7. *Connections to be made between the different modules for exercise 10 – reverb*

3.3.1.6. *Reaktor*

In the following rack, we will be able to test three reverb blocks and a delay/echo block from the Native Instruments user library:

– Air Reverb, developed by Jonathan Tremblay;

– Pimped Rounds Reverb, developed by Thala Estra;

– SchepperVerb III, developed by Dieter Zobel;

– dEIA, developed by Charles Capsis IV.

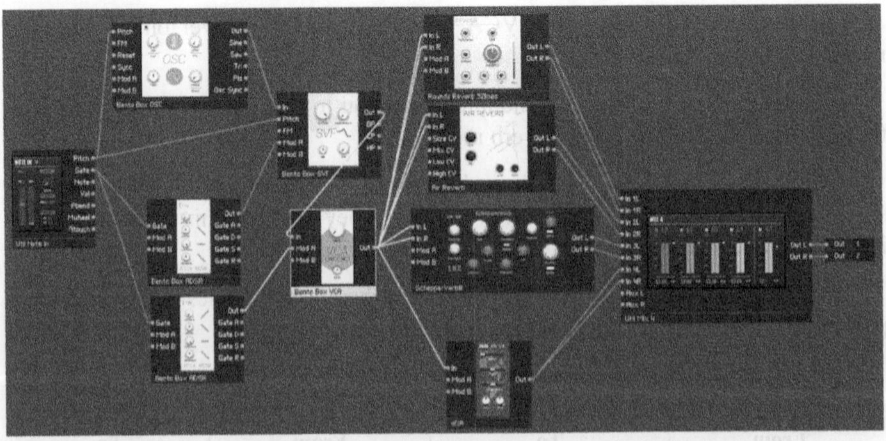

Figure 3.28. *Three reverb modules in a Reaktor rack. For a color version of this figure, see www.iste.co.uk/reveillac/synthesizers2.zip*

These blocks are sophisticated and have different sound renderings. Note that they are placed behind the VCA block, and the first three are stereophonic.

The MIX 4 block ensures muting and possible mixing to the audio output.

The remaining blocks form a standard configuration: a VCO, a VCF, a VCA and two envelope generators, as we have already seen.

Figure 3.29. *Rack in Panel mode. For a color version of this figure, see www.iste.co.uk/reveillac/synthesizers2.zip*

Instructions and comments are as follows:

– the output of the VCA block is connected to each of the IN L and IN R inputs of the REVERB, AIR REVERB and SCHEPPERVERB blocks;

– the output of the VCA block is connected to the single IN input of the dEIA block;

– the outputs of the OUT L and OUT R blocks of each of the three blocks REVERB, AIR REVERB and SCHEPPERVERB are connected to inputs 1, 2 and 3, left and right of the MIX 4 block;

– the OUT output of the dEIA block is connected to the two inputs IN 4L and IN 4R of the MIX 4 block;

– the OUT L and OUT R outputs of the MIX 4 block are connected to the AUDIO OUT output blocks;

– by checking each of the four M indicators of the MIX 4 block, you can disable the signal output from the reverb and delay blocks;

– each of the four faders on the left of the MIX 4 block allows you to vary the gain of each input. The right fader affects the general gain;

– try the settings of the different blocks independently. You will notice the difference made by the dEIA block dedicated to delay and echo rather than reverberation;

– note that with this wiring, in parallel with the four effect blocks, you can mix the different reverberations and echo according to your tastes.

3.3.2. *Exercise 11 – chorus*

This effect is widely used to enrich and thicken a sound; it gets its name from the way in which a chorus of people sing sounds in unison, which differs from a single singer. The same goes for a group of violins playing the same note or on a piano, whereby each hammer strikes two or three strings tuned in unison. Slight fluctuations, due to a reasonably short delay of 20–40 ms compared to the original signal, create interference between the emitted sound waves and provide a sound rendering that is always different over time and often unpredictable.

3.3.2.1. *Max/MSP*

In this version of the Max/MSP chorus effect, variations of the chorus effect will be processed randomly by overlaying slightly modified copies of the original signal.

The `tapin~` and `tapout~` objects create a retroactive delay line, and the `rand~` object generates small, continuous fluctuations in delay time that are linearly interpolated with the original signal. The value passed to the `rand~` object defines the temporal rhythm of fluctuations.

This is not the only solution for creating the chorus effect; there are others that we will see a little later (asynchronous vibrato, noise, modification of the fundamental, etc.).

In this patcher, the chorus engine is managed by a `ChorusEffect` subpatch called from the main patch.

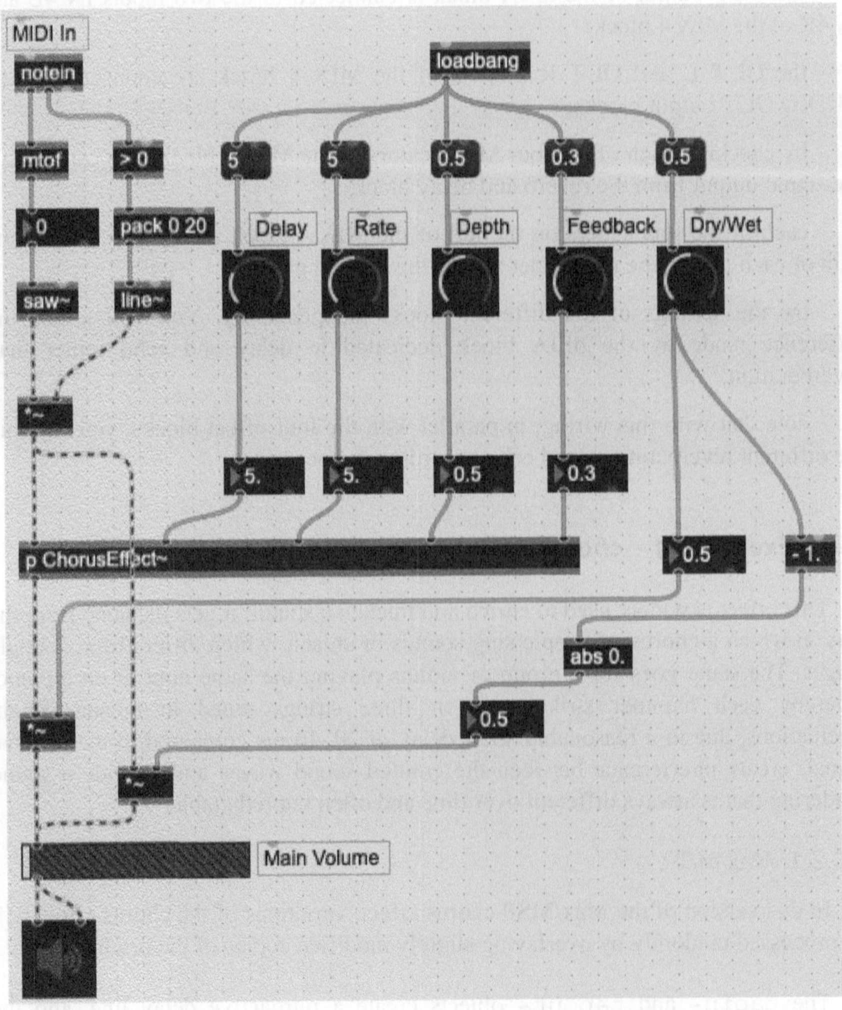

Figure 3.30. *A patcher for the chorus effect with Max/MSP*

Instructions and comments are as follows:

– the seven objects, top and left, placed behind the notein~ object have already been discussed (see section 2.2.3, exercise 4b);

– the loadbang object preloads the values contained in the five following message objects when launching the patch;

– the five dial objects adjust the various chorus parameters;

– the dial Delay object has a size of 21 and a minimum value of 0;

– the dial Rate object has a size of 10 and a minimum decimal value of 0;

– the dial Depth, dial Feedback and dial Dry/Wet objects have a size of 1 and their minimum decimal value is 0;

– the five flonum objects following the dial object allow you to control the values chosen by the user;

– the p ChorusEffect~ object calls the ChorusEffect subpatch;

– the - 1 and abs 0. objects manage the balance between the original signal and the processed signal;

– the remaining *~ objects combine the original and processed signals;

– the main object manages the main volume;

– the ezdac~ object ensures audio output;

– the subpatch has five inlet objects (input) and an outlet~ object (output);

– the tapin~ 20 object is a default 20 ms delay line that retains part of the original signal to transmit to the tapout~ object;

– the rand~ object acts as a random low-frequency oscillator that modulates the signal present within the delay line;

– the tapout~ object restores the part of the original signal, modulated by the random low-frequency oscillator to create the chorus effect;

– the inlet 1 object retrieves the value of the dial Delay object and controls the delay from 0 to 20 ms;

– the inlet 2 object retrieves the value of the dial Rate and controls the modulation rate;

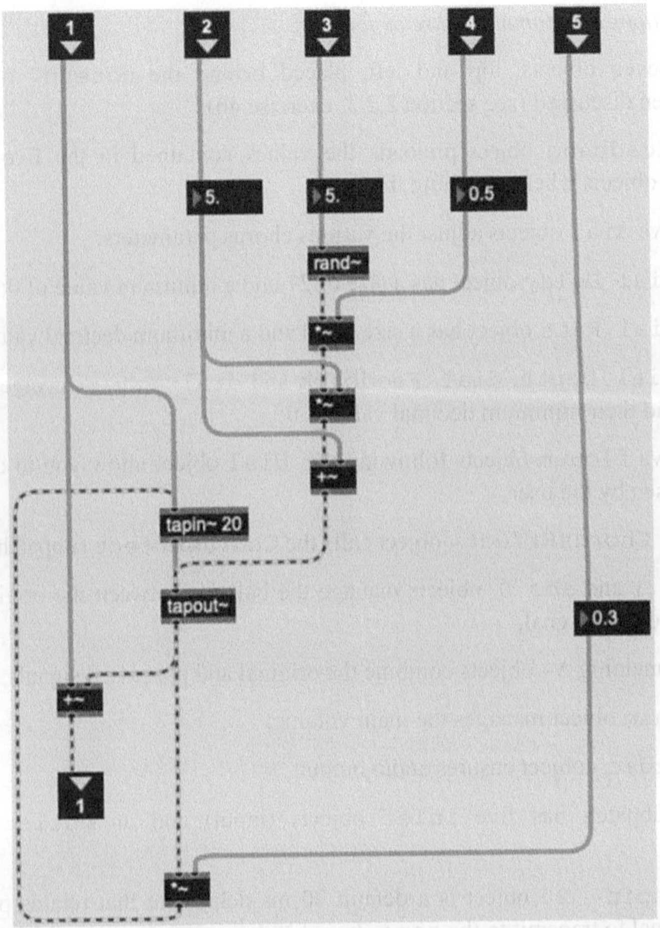

Figure 3.31. *The ChorusEffect subpatch of the chorus effect*

– the `inlet 3` retrieves the value of the `dial Depth` and controls the depth of the chorus effect;

– the `inlet 4` retrieves the value of the `dial Feedback` and controls the feedback level by feeding the signal back into the `tapin~`;

– the `inlet 5` retrieves the value of the `Dry/Wet` object, which balances the original and the processed signals;

– the 4 `flonum` objects show the different values transmitted by the `dial` objects;

– the two *~ objects and the +~ object combine and add the different parameters for the tapout ~ object;

– the +~ object placed in front of the output sums the original signal and the processed signal;

– try the different settings and note their influence on the sound reproduction. You must find the right compromise so that the audio signal remains as natural as possible, although it is sometimes interesting to force the line;

– remember to set the overall volume to the correct level.

3.3.2.2. *Pure Data*

Unlike the solution adopted with Max/MSP, in this Pure Data exercise, we will use a sinusoidal oscillator to generate the variations necessary for a chorus effect.

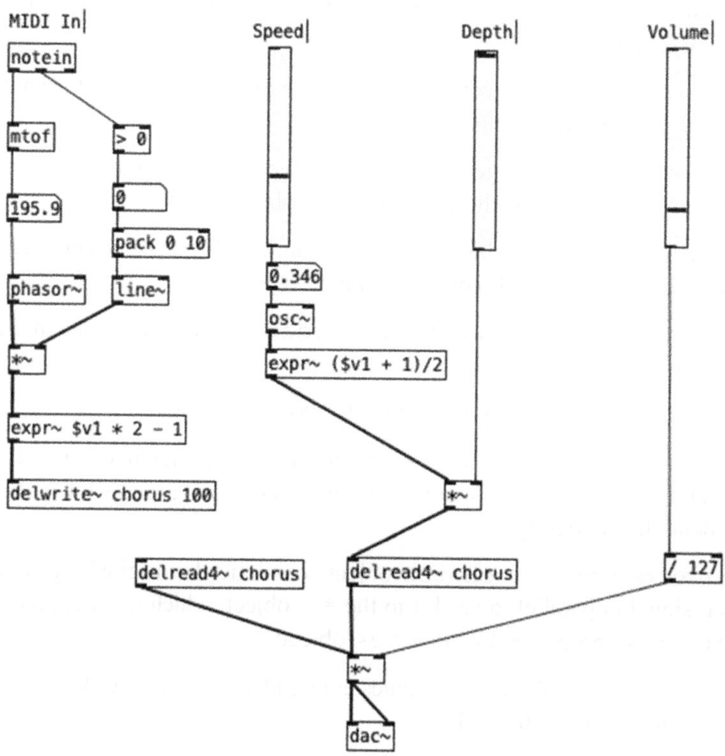

Figure 3.32. *A patch for the chorus effect with Pure Data*

For this effect, the settings available to the user will be speed, depth and general volume.

We will use a delay line to keep part of the signal, which will be read back after being modulated. Then, the original signal and the modulated signal will be associated.

Instructions and comments are as follows:

– the objects on the left provide the audio signal from the MIDI keyboard. The signal has a unipolar sawtooth waveform generated by the `phasor~` object. It is amplified and transformed into a bipolar signal by the `expr~ $v1 * 2 - 1` object, then sent to the line object with a delay of 100 ms, `delwrite~ chorus 100`;

– the three `slider Speed`, `Depth`, and `Volume` objects ensure that settings are available to the user;

– the `slider Speed` object manages a range of linear values between 0 and 1. It defines the speed of the chorus;

– the `slider Depth` object manages a range of linear values between 0 and 10. It defines the depth of the chorus effect by amplifying it;

– the `slider Volume` object manages a range of linear values between 0 and 127. It defines the general audio output volume;

– the `number` following the `slider Speed` object shows the speed value chosen by the user;

– the `osc~` object generates the sinusoidal modulation signal;

– the `expr~ ($v1 + 1) /2` object shapes the signal coming from the oscillator and transforms it into a unipolar signal in order to process the audio signal from the delay line correctly;

– the two `delread4~ chorus` objects combine the original signal and the modulated signal in parallel to send it to the `*~` object, which manages the volume of the audio output processed by the `dac~` object;

– the `/ 127` object divides the values emitted by the `slider Volume` object to have a volume between 0 and 1;

– try the different speed and depth settings of the effect and listen to the changes they make to the sound reproduction;

– if the speed setting is zero, no chorus effect is heard.

NOTE.– You will notice that, in this patch, the original signal systematically passes through a delay line, unlike many solutions, which take the signal before the line to mix it with the processed signal. This makes the left part of the patch (between `notein` and `delwrite~ chorus 100`) not seem attached to other objects.

3.3.2.3. *VCV Rack*

This rack has an identical configuration to the previous VCV Rack exercise (see section 3.3.1.5). Still, the reverb modules have been replaced by chorus modules, one of which is associated with an LFO.

The MUTES module ensures the signal is routed to the three chorus modules, the fourth input allowing the signal to be diverted (bypass) directly to the audio output via the VCA MIX mixing module.

The chorus modules selected are as follows:

– Vult Jorus, developed by Leonardo Laguna Ruiz (freeware);

– NYSTHI Stereo Chorus, developed by Antonio Tuzzi and Nysthi (freeware);

– Autodafe Chorus, developed by Autodafe (freeware).

NOTE.– For more details, consult the VCV Rack library. You can access much information and often a user manual by choosing the module considered.

Instructions and comments are as follows:

– create a new rack and add the following modules: MIDI-CV, OCT, VCO, 2 × ADSR, VCF, MUTES, VCA MIX, AUDIO, LFO and the three chorus modules, or repeat the previous exercise and delete the four reverb modules;

– connect the inputs–outputs of the different modules, as shown in Figure 3.33 and Table 3.8. The envelope generators are differentiated according to their destination, VCF or VCA, on the left and right in Figure 3.33;

– do not forget to adjust the VCA volume and output level on the AUDIO module;

– the buttons on the MUTES module allow you to open (gray) or close (white) the signal passage to one of the chorus modules or the VCA MIX module for input/output no. 4 (bypass);

– the sliders of the VCA MIX module allow you to adjust the volume of the different signals. They can also mix several sources if the passages have been opened from the MUTES module;

– the FREQ knob on the LFO module varies the movement of the NYSTHI chorus. You can try other controls (MOD FREQ, MOD DEPTH, FEEDBACK, etc.) and choose a different output on the LFO module (TRI, SAW or SQR) to listen to the changes. Do not forget to adjust the controls on the NYSTEREOCHORUS module;

– test the different chorus modules with different settings to hear the changes in sound reproduction.

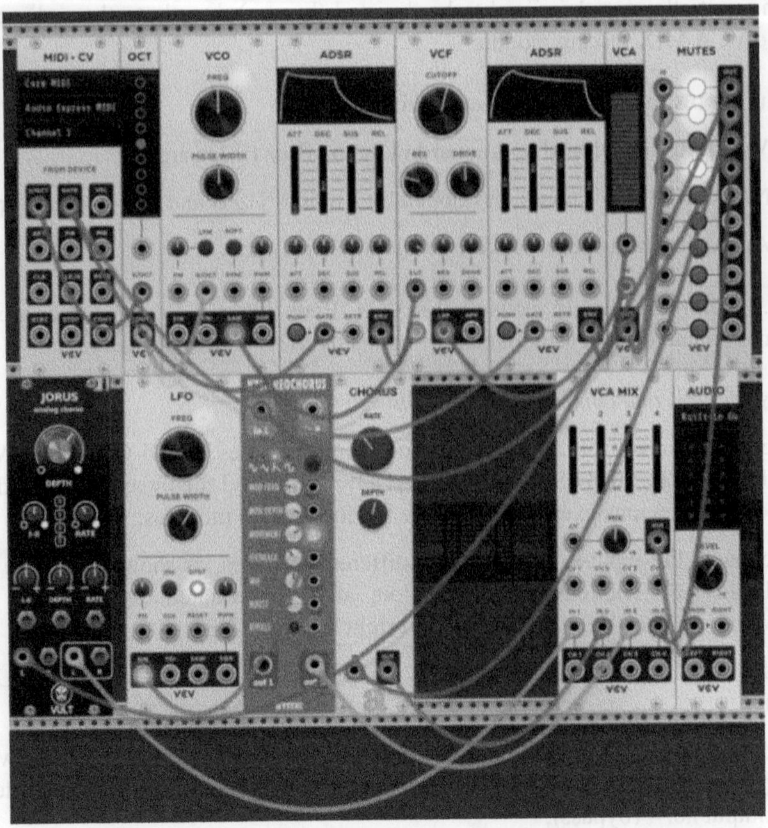

Figure 3.33. *Chorus and VCV Rack. For a color version of this figure, see www.iste.co.uk/reveillac/synthesizers2.zip*

From	To	From	To
MIDI-CV–V/OCT	OCT–V/OCT	**VCA–OUT**	MUTES–IN 4
MIDI-CV–GATE	ADSR (VCF) – GATE	**MUTES–OUT 1**	JORUS–L (In)
MIDI-CV–GATE	ADSR (VCA)–GATE	**MUTES–OUT 2**	NYSTEREOCHORUS–IN L
OCT–OUT	VCO–V/OCT	**MUTES–OUT 3**	CHORUS (Autodafe)–IN
VCO–SAW	VCF–IN	**MUTES–OUT 4**	VCA MIX–IN 4
ADSR (VCF) – ENV	VCF–CUT	**JORUS–L (Out)**	VCA MIX–IN 1
VCF–LPF	VCA–IN	**LFO–SIN**	NYSTEREOCHORUS–MOVEMENT
ADSR (VCA)–ENV	VCA–CONTROL	**NYSTEREOCHORUS–OUT L**	VCA MIX–IN 2
VCA–OUT	MUTES–IN 1	**CHORUS (Autodafe)–OUT**	VCA MIX–IN 3
VCA–OUT	MUTES–IN 2	**VCA MIX–MIX**	AUDIO–L/MON
VCA–OUT	MUTES–IN 3	–	–

Table 3.8. *Connections to be made on the VCV Rack for exercise 11 – chorus*

3.3.3. *Exercise 12 – flanger*

The flanger changes the sound at random intervals. A constantly varying short delay time (2–15 ms) is applied to the signal, which is then combined with the original signal. It widens the sound and makes it livelier.

It is more extreme than the chorus and creates a metallic effect. Musicians often refer to flanger as the sound of an airplane engine.

Several parameters are generally available: modulation frequency, depth and dry/wet balance. The modulation frequency is the essential parameter, and it must be carefully controlled; otherwise, you will obtain a particularly distorted and unusable sound rendering.

There are several methods to create an artificial flanger:

– Additive: the delayed signal is added;

– Subtractive: the delayed signal is added inversely to the shape of the curve of the first signal;

– Through-Zero: a delay is applied to the original signal so that the modulated signal comes before it.

3.3.3.1. *Max/MSP – Flanger 1*

In Figure 3.34, you will find the first version of a Max/MSP patcher offering a flanger effect. This is based on the classic use of `tapin~` and `tapout~` objects for delay management, associated with a sinusoidal oscillator to generate the modulation.

Unlike the previous exercises and examples, the dry/wet mixture has been assigned to the `balance~` subpatch object, thus providing a new programming solution.

The other objects are placed in the simple configuration we have already seen.

With the `tapin~` and `tapout~` objects, delay management is based on a buffer size and a delay time in milliseconds.

Instructions and comments are as follows:

– we will not discuss the patcher and how it generates the audio signal from the MIDI keyboard, as this is now standard;

– the four `flonum Feedback`, `Rate`, `Depth` and `Delay` objects control the flanger parameters;

– the `flonum Feedback` and `flonum Depth` objects are limited to a range of decimal values between 0 and 1;

– the `flonum Rate` object covers a range of decimal values from 0 to 20;

– the `flonum Delay` object covers a range of decimal values from 0 to 200;

– the `cycle~` object generates the sinusoidal modulation frequency, by default 3 Hz;

– the two `*~` objects and `+~` object combine and sum the modulation rate, depth and delay. By default, the depth is 0.1, and the delay is 20 ms;

– the `*~` object that follows the `flonum Feedback` object ensures a feedback loop;

– the `tapin~` object is the input delay line with a maximum delay of 250 ms by default, and the `tapout~` object is the output delay line;

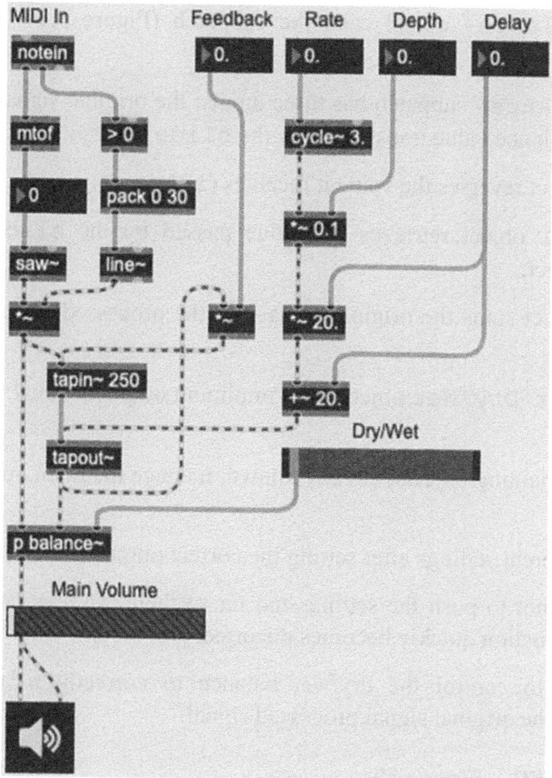

Figure 3.34. *A patcher for the flanger effect*

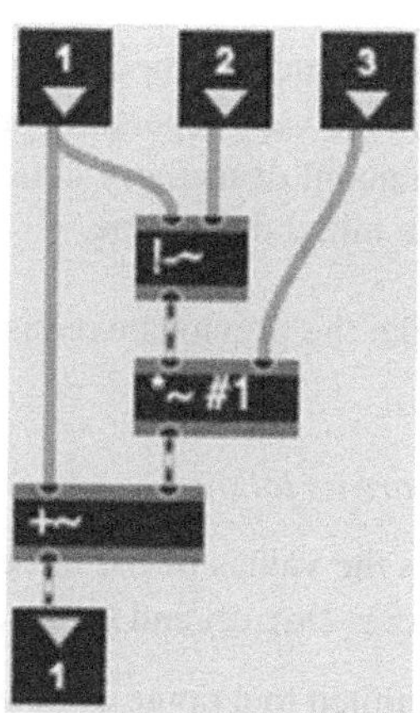

Figure 3.35. *The balance~ subpatch*

– the p balance~ object calls the subpatch (Figure 3.35) of the dry/wet balance;

– the p balance~ subpatch has three inputs: the original signal, the processed signal and the balance value transmitted by the slider Dry/Wet;

– the !~ object reverses the value it receives (inlet 2);

– the *~ #1 object retrieves the value passed by the slider Dry/Wet (inlet 3) object;

– the +~ object sums the original signal and the processed signal to control the balance;

– the slider Dry/Wet object has a minimum output decimal value of 0 and a size of 1;

– the two remaining objects, already known, manage the main volume and audio output;

– try the different settings after setting the correct output volume;

– be careful not to push the settings too far to their extreme values; otherwise, the sound reproduction quickly becomes distorted;

– remember to control the dry/wet balance to correctly balance the sound reproduction of the original signal/processed signal.

3.3.3.2. *Max/MSP – Flanger 2*

In this second exercise, the Max/MSP simulates the flanger effect. Here, the tapin~ and tapout~ objects have been replaced by a delay~ object.

The delay~ manages input values defined as the number of samples (default 512), which causes the delay time to vary depending on the sampling frequency. The first argument determines the amount of memory allocated to the delay line, and the second argument determines the initial delay time.

The rest of the patcher is like the previous exercise; however, no feedback loop is present.

Instructions and comments are as follows:

– the loadbang~ preloads the values of the three message objects (2, 30 and 50) in the three number~, Rate, Depth and Delay objects;

– the number~ Rate is limited to a range of decimal values between 0 and 20;

– the `number~ Depth` is limited to a range of values between 0 and 500;

– the `number~ Delay` is limited to a range of decimal values between 0 and 200;

– the `cycle~` creates modulation;

– the `delay~` creates a delay;

– the `p balance~` subpatch is identical to that of the previous exercise; it manages the dry/wet balance;

– the `slider Dry/Wet` has a minimum output decimal value of 0 and a size of 1;

– all other objects have already been encountered;

– try the different settings and choose the right dry/wet balance.

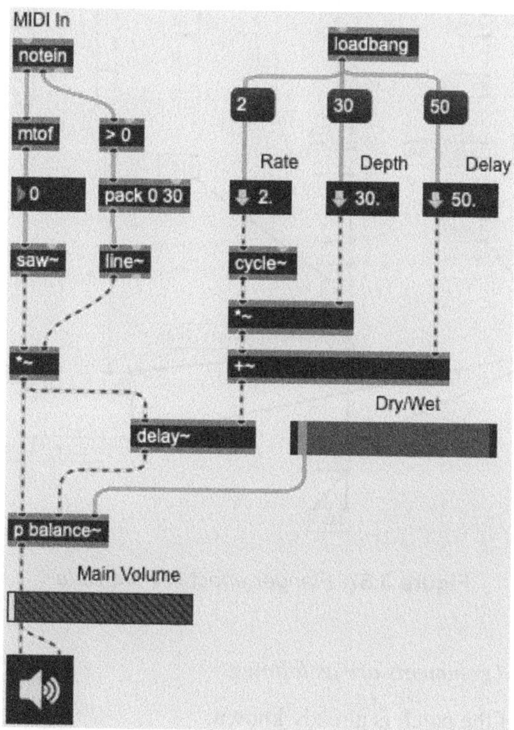

Figure 3.36. *A patcher using the delay~ object for the flanger effect*

3.3.3.3. *Pure Data*

For this Pure Data patch, we use a similar configuration to that already created for the chorus during the previous exercise.

Here, the original signal is shared directly with the processed signal, and the sinusoidal modulation signal shape is different. On the other hand, feedback occurs by reinjecting the modulated signal, read in the delay line, toward the writing of this line.

The rest of the patch uses the typical sequence of objects, already encountered in the previous exercises.

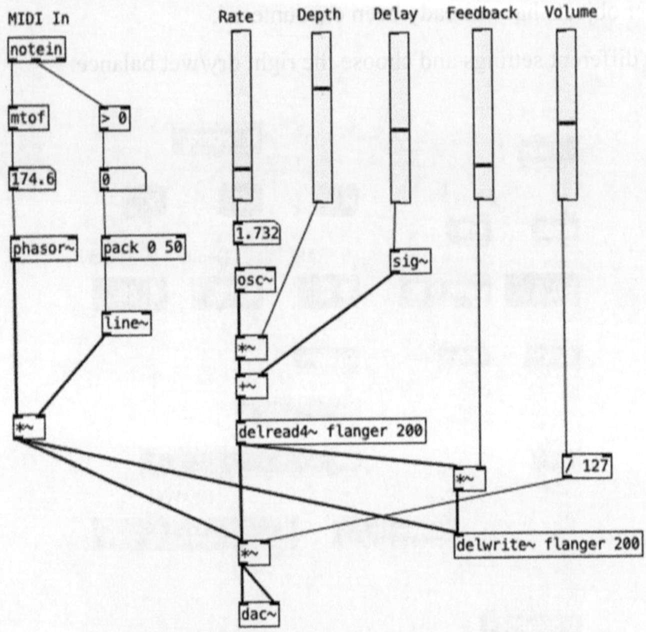

Figure 3.37. *Flanger effect in Pure Data*

Instructions and comments are as follows:

– the left part of the patch is already known;

– the five `slider` objects manage the different flanger parameters;

– the `slider Rate` object covers a range of linear values between 0 and 10;

– the `slider Depth` object covers a range of linear values between 0 and 5;

– the `slider Delay` object covers a range of linear values between 0 and 200;

– the `slider Feedback` object covers a range of linear values between 0 and 1;

– the `slider Volume` object covers a range of linear values between 0 and 127;

– the `osc~` object generates the sinusoidal modulation signal;

– the `sig~` object converts the numeric value sent by the `slider Delay` object into an audio signal;

– the two `*~` and `+~` objects multiply and sum the different parameters destined for the delay line;

– the `delread4~ flanger 200` object is the delay line available for reading with a duration of 200 ms by default;

– the `delwrite~ flanger 200` object is the delay line in which the modulated signal is written;

– the `*~` object behind the `slider Feedback` object considers the chosen level of feedback;

– the other objects have already been discussed; they ensure volume and audio output control, mixing the original and processed signals;

– try the different parameters to see their influence on the sound reproduction;

– avoid forcing the values too much so that you have an easy-to-use audio signal.

3.3.3.4. *VCV Rack*

The VCV Rack has a few flanger modules; in this exercise, the Sckitam WaveguideDelay module was selected for its very interesting performance.

Philippe Salembier developed this module under the GPL 3.0 license.

Thanks to its features and multiple connection points, it responds to effects ranging from delay to flanger to chorus. Initially, it was designed for Karplus–Strong synthesis, but it can easily be diverted from its primary objective.

Karplus–Strong synthesis, named after its two inventors, is a set of techniques that make it possible to synthesize the sound of plucked string and percussion instruments realistically. This is an algorithm referred to as physical modeling synthesis.

NOTE.– For more details, you can access additional information from the VCV Rack library, and the user manual by choosing the Sckitam WaveguideDelay module.

Figure 3.38. *Flanger for the VCV Rack. For a color version of this figure, see www.iste.co.uk/reveillac/synthesizers2.zip*

Instructions and comments are as follows:

– repeat the previous exercise and delete the modules that do not exist in this exercise, or create a new rack and add the following modules: MIDI-CV, OCT, VCO, 2 × ADSR, VCF, VCA, LFO, VCA MIX, AUDIO and WAVEGUIDE DELAY;

– connect the inputs–outputs of the different modules as shown in Figure 3.38 and Table 3.9. The envelope generators are differentiated according to their destination, VCF or VCA, on the left and on the right in Figure 3.38;

– in the WAVEGUIDE DELAY module, the DELAY switch sets the length of the delay, the DADJ switch finely adjusts the delay time, the upper PK POS switch manages the delay capture position in the delay buffer (see next point), the lower PK POS button manages the modulation possible in CV mode of the capture position and the FBACK button controls the feedback level, between 0 and 0.999;

– in this same module, the available connection points are IN for the delay input, OUT for the main delay output, FBACK for the entry point of a feedback signal, PICKUP for the output of the signal taken from the delay buffer (0 for the delay input and 1 for the output, adjustable by a button), DELAY for managing delay time in 1V/Oct mode, and PK POS for modulation input, in CV mode, of the delay capture point in the delay buffer;

From	To	From	To
MIDI-CV–V/OCT	OCT–V/OCT	**ADSR (VCA)–ENV**	VCA–CONTROL
MIDI-CV–GATE	ADSR (VCF) –GATE	**VCA–OUT**	VCA MIX–IN 1
MIDI-CV–GATE	ADSR (VCA)–GATE	**VCA–OUT**	WAVEGUIDE DELAY–IN
OCT–OUT	VCO–V/OCT	**LFO–SIN**	WAVEGUIDE DELAY–DELAY
VCO–SAW	VCF–IN	**WAVEGUIDE DELAY–OUT**	VCA MIX–IN 2
ADSR (VCF) – ENV	VCF–CUT	**WAVEGUIDE DELAY–FBACK**	WAVEGUIDE DELAY–PICKUP
VCF–LPF	VCA–IN	**VCA MIX–MIX**	AUDIO–L/MON

Table 3.9. *Connections to make for exercise 12 – flanger*

– try the different settings and observe the influence of the LFO on the sound reproduction;

– you can also connect the LFO outputs to the PK POS connection point to influence the position of the capture point in the delay buffer. Change the influence of the input signal by turning the knob above;

– using the VCA MIX module, you can manage the dry/wet balance of the effect via the IN 1 and IN 2 faders (max IN 1 and min IN 2 for dry and vice versa for wet).

3.3.4. *Exercise 13 – phaser*

The phaser adds the copy of a signal filtered using an all-pass filter (or phase-shifting filter) to an original signal. The objective is to modify the latter by generating interference, which cancels or reinforces the phase according to the determined frequencies within the filter. The coefficients may vary but we consider that values between 0.88 and 0.98 are correct. There are many solutions and variations to achieve this result, including the use of 12 or more filters in series. The audio spectrum is cyclically modified by a low-frequency oscillator, which creates a whirling effect.

Unlike the flanger, which is based on a comb filter varying over time, the pitch is not imposed for the phaser.

The main settings of a phaser are the sweep frequency and the range of all-pass filter coefficients. Feedback can be installed within a phaser by reinjecting part of the signal into the all-pass filters. Under no circumstances should the delay time of the filters be varied; otherwise, reverberation could appear.

3.3.4.1. *Max/MSP*

In this version of a phaser for Max/MSP, we will use a device using four all-pass filters. A feedback loop will provide feedback.

The user can adjust the speed, feedback, dry/wet balance and main volume.

Instructions and comments are as follows:

– the note generation part in the top left is standard; we will not discuss this again;

– the `loadbang` object preloads the default values of feedback, speed and dry/wet balance, that is, 0.5, 1.5 and 0.5, respectively, from the `message` objects;

– the three `dial` objects adjust the various parameters;

– the `dial Feedback` object has a size of 1 and a minimum decimal value of 0;

– the `dial Speed` object has a size of 10 and a minimum decimal value of 0;

– the `dial Dry/Wet` object has a size of 1 and a minimum decimal value of 0;

– the `send~ f` and `receive~ f` objects ensure the feedback loop;

– the four `allpass~` objects ensure the phase shift of the signal;

– the `phasor~`, `cos~`, `*~ 0.5`, `+~ 0.5`, `*~ 0.1` and `+~ 0.88` objects shape the sawtooth oscillator signal, making it sinusoidal, then unipolar, and assign it the correct amplitude;

– the `number~` object visualizes the variations of the oscillator;

– the `- 1.`, `abs 0.` objects followed by two `*~` objects manage the dry/wet balance;

– the remaining objects take care of managing the general volume and audio output;

– try the different settings; be careful with the feedback setting to avoid a somewhat inconsistent sound reproduction.

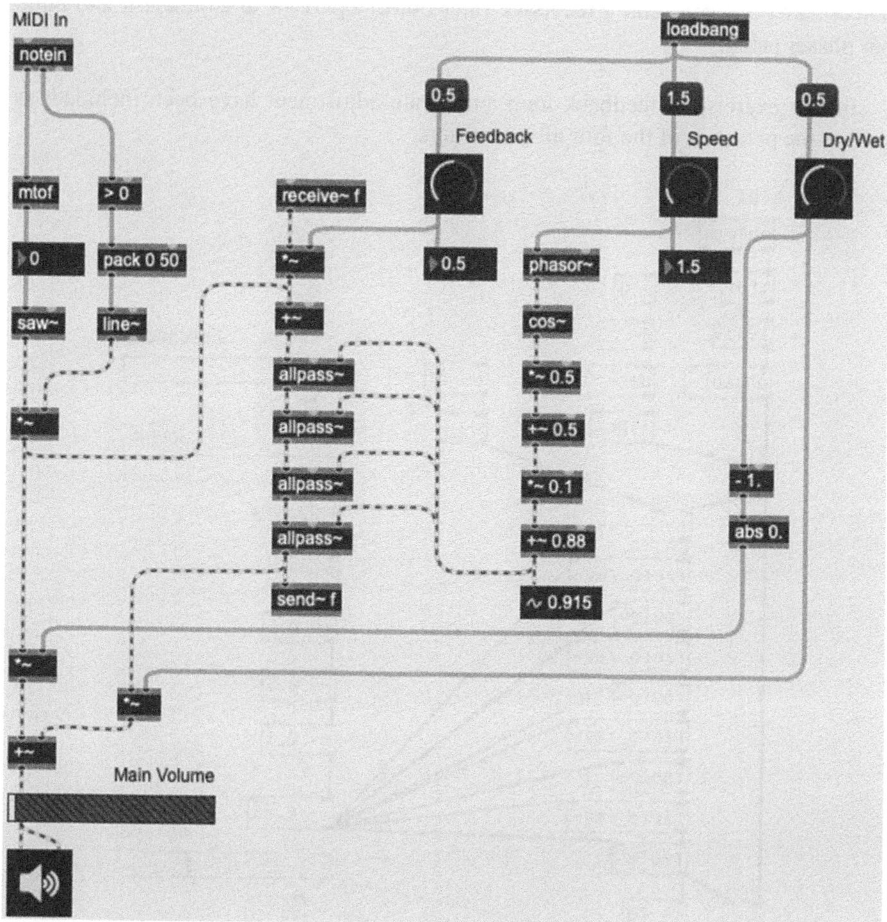

Figure 3.39. *Phaser with Max/MSP*

3.3.4.2. *Pure Data – Phaser 1*

Below, we have two different techniques for generating a phaser effect. The first exercise is based on the use of a four-stage filter made up of all-pass filters to obtain a phase shift, and the second on the phase-shifted modulation of the same signal placed in two delay lines.

Pure Data does not have an object that simulates an all-pass filter, so we will use a solution conceived by Miller Puckette, which associates a non-recursive filter object `rzero_rev~` and a recursive filter object `rpole~` to simulate it and build our phaser patch.

In this exercise, a feedback loop and a rate adjustment have been included, as well as the presence of the four all-pass filters.

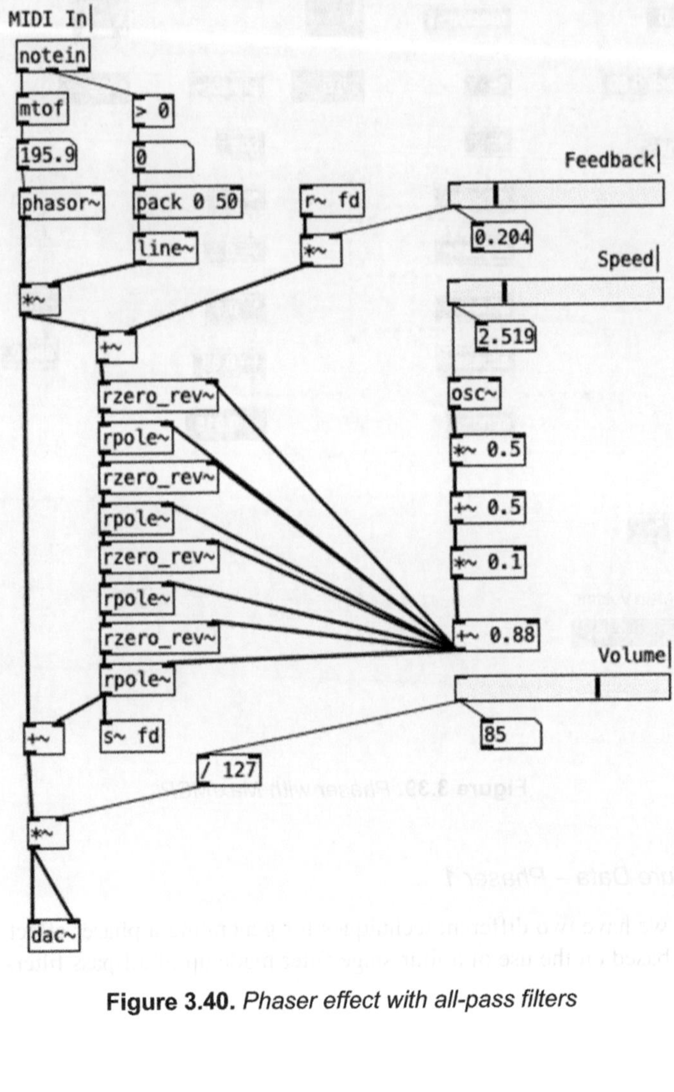

Figure 3.40. *Phaser effect with all-pass filters*

Instructions and comments are as follows:

– the upper left part of the patch is already known to you;

– the right part contains the different settings managed by three `slider` objects;

– the `slider Feedback` object handles a range of linear values between 0 and 1;

– the `slider Speed` object handles a range of linear values between 0 and 10;

– the `slider Volume` object handles a range of linear values between 0 and 127;

– the `osc~` object represents the low-frequency oscillator;

– the `*~ 0.5`, `+~ 0.5`, `*~ 0.1`, `*~ 0.88` objects conform the oscillator signal, make it unipolar, and assign it the correct amplitude coefficient;

– the four pairs of `rzero_rev~` and `rpole~` objects form the four stages of all-pass filters;

– the `s~ fd` and `r~ fd` objects transmit and receive the processed signal to ensure the creation of the feedback loop;

– the remaining objects provide mixing, volume adjustment and audio output;

– try the different settings and listen to the sound reproduction.

3.3.4.3. *Pure Data – Phaser 2*

In this second exercise, the phaser effect will be generated with the help of two delay lines whose signals will be out of phase with each other.

We will find five user-adjustable parameters: speed, depth, feedback, dry/wet balance and general volume.

Instructions and comments are as follows:

– the objects in the upper left part generate a note sent to the audio output and the two delay lines: the `delwrite~ phase 1 200` and `delwrite~ phase 2 200` objects;

– the `phasor~` object placed behind the `slider Speed` object creates the modulation;

– the `slider Speed` object handles a range of linear values between 0 and 1;

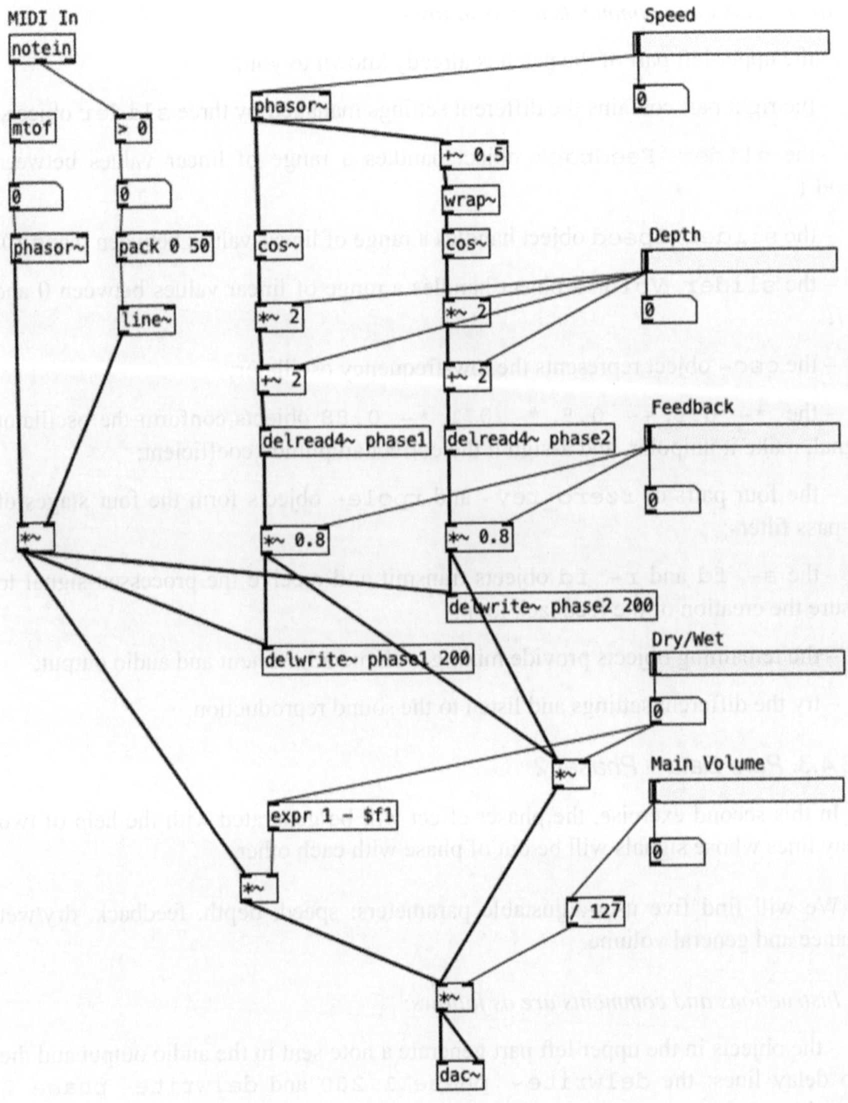

Figure 3.41. *Phaser effect with two delay lines*

– the two `cos~` objects transform the sawtooth signal of the `phasor~` object into a sinusoidal signal;

– the two `+~ 0.5` and `wrap~` objects modify the phase of the modulation signal;

– the two object pairs `*~ 2` and `+~ 2` modify the amplitude of the signals and make them unipolar according to the value defined by the `slider Depth` object, default 2;

– the `slider Depth` object handles a range of linear values between 0 and 4;

– the two `delread4~ phase 1` and `delread 4~ phase 2` objects read the contents of the delay lines parameterized by the phase-shifted modulation signals;

– the two `*~ 0.8` objects ensure the feedback loop by returning part of the signal to the two delay lines and transferring the processed signals to the audio output;

– the `slider Feedback` object controls the feedback. It manages a range of linear values between – 0.99 and 0.99;

– the two `delwrite~ phase 1` and `delwrite~ phase 2` objects receive the original signal and write it into a delay line of duration equal to 200 ms;

– the `expr 1 - $f1` object, the `slider Dry/Wet` object and two `*~` objects recover the original signal and the two signals coming from the delay lines to mix them and balance the original signal and processed signals;

– the `slider Dry/Wet` object covers a range of linear values between 0 and 1;

– the `slider Main Volume` object, the `/ 127` object and the `*~` object placed right in front of the `dac~` object ensure overall volume adjustment before outputting the audio;

– each `slider` object is followed by a `number` object allowing the chosen value to be displayed;

– try the different settings and listen to the influence of depth, speed and feedback;

– do not forget to fine-tune the dry/wet balance and volume.

3.3.4.4. *VCV Rack*

The AS Phaser Fx, a single phaser module developed by Alfredo Santamaria, has been integrated into this rack.

This module is a classic phaser and provides an excellent phaser effect. A bypass mode allows you to disable the passage through the effect. All three parameters, FX, Rate and Feedback can be controlled in CV mode if desired.

Figure 3.42. *A rack integrating the AS Phaser Fx module. For a color version of this figure, see www.iste.co.uk/reveillac/synthesizers2.zip*

Instructions and comments are as follows:

– this rack has a classic configuration: VCO, VCF, 2 × EG and a VCA;

– the AS Phaser Fx module is wired to the VCA output just before the AUDIO module (see Figure 3.42);

– try the different phaser settings; the sound reproduction is very realistic with progressive modulation and phase shift, and the feedback doing its job correctly.

3.4. Conclusion

This chapter explored the different functions and sound effects found on many synthesizers.

Hardware synthesizers are generally simple to implement and use. However, programming them in a visual programming language like Max/MSP or Pure Data is very different. You must have good knowledge of the principles of each of them to design a solution that best matches their characteristics.

With both Max/MSP and Pure Data, the solutions to the exercises presented previously are not the only ones, and there are many other equally effective solutions. Do not hesitate to consult the reference section at the end of the book if you want to explore this further.

4

Duophony, Paraphony and Polyphony

Although already presented in Chapter 3 of Volume 1, let us recap the different modes of managing voices in a synthesizer.

A polyphonic synthesizer can play multiple independent notes to create chords, an accompaniment, bass lines, etc. In the early days of electronic music, most instruments were monodic; only one note could be played at a time.

The number of oscillators is not the determining factor of polyphony; many monodic synthesizers have several oscillators, such as the Minimoog, the ARP 2600 and the Korg MS20.

Besides monophonic synthesizers, there are duophonic or paraphonic ones, which are more versatile.

A duophonic synthesizer like the ARP Odyssey can independently play two notes at a time, which requires the machine to have at least two oscillators and a keyboard capable of managing two control voltages.

Paraphonic machines like the Korg Poly800 or Arturia MatrixBrute are designed to play multiple pitches simultaneously using multiple oscillators but sharing the same filter. Some monodic synthesizers, such as the ARP 2600, can be transformed or rewired – when modular or semi-modular – into paraphonic ones.

So, the question is, how many oscillators must a synthesizer have to be polyphonic? In simple terms, only 12 are needed, one for each note in the scale. Additional notes are created by dividing the signal from each of the basic oscillators. To produce a note one octave lower, divide the frequency by two. The Polymoog and the Korg PS3300 are based on this technique, like many electronic organs from the 1980s.

In this chapter, we will study each of these voice management methods through several exercises.

You will find in Table 4.1, for each exercise, the hardware and software chosen and the sections where you can find them.

<table>
<tr><th>Exercise</th><th>Hardware/software</th><th>Section</th></tr>
<tr><td rowspan="5">No. 14. Duophony and paraphony</td><td>Behringer 2600 (ARP2600)</td><td>4.1.1</td></tr>
<tr><td>Novation Bass Station II</td><td>4.1.2</td></tr>
<tr><td rowspan="2">Behringer Neutron</td><td>4.1.3.1</td></tr>
<tr><td>4.1.3.2</td></tr>
<tr><td>Arturia MatrixBrute</td><td>4.1.4</td></tr>
<tr><td rowspan="8">No. 15. Polyphony</td><td rowspan="3">Max/MSP</td><td>4.2.1.1</td></tr>
<tr><td>4.2.1.2</td></tr>
<tr><td>4.2.1.3</td></tr>
<tr><td rowspan="2">Pure Data</td><td>4.2.2.1</td></tr>
<tr><td>4.2.2.2</td></tr>
<tr><td rowspan="2">VCV Rack</td><td>4.2.3.1</td></tr>
<tr><td>4.2.3.2</td></tr>
<tr><td>Reaktor</td><td>4.2.4</td></tr>
</table>

Table 4.1. *Summary of the different exercises*

4.1. Exercise 14 – duophony and paraphony

In this exercise, we will see how to use duophony or paraphony with four different synthesizers: the Behringer 2600 (ARP 2600), the Novation Bass Station II, the Behringer Neutron and the Arturia MatrixBrute.

Playing a melody on duophonic or paraphonic equipment requires skill to bring out the multivocal aspect, which often lends itself instead to slow playing or subtle timbre changes spanning very long periods.

4.1.1. *Behringer 2600 (ARP 2600)*

The ARP 2600 offers a possible duophony mode; here, it is shown using a patch.

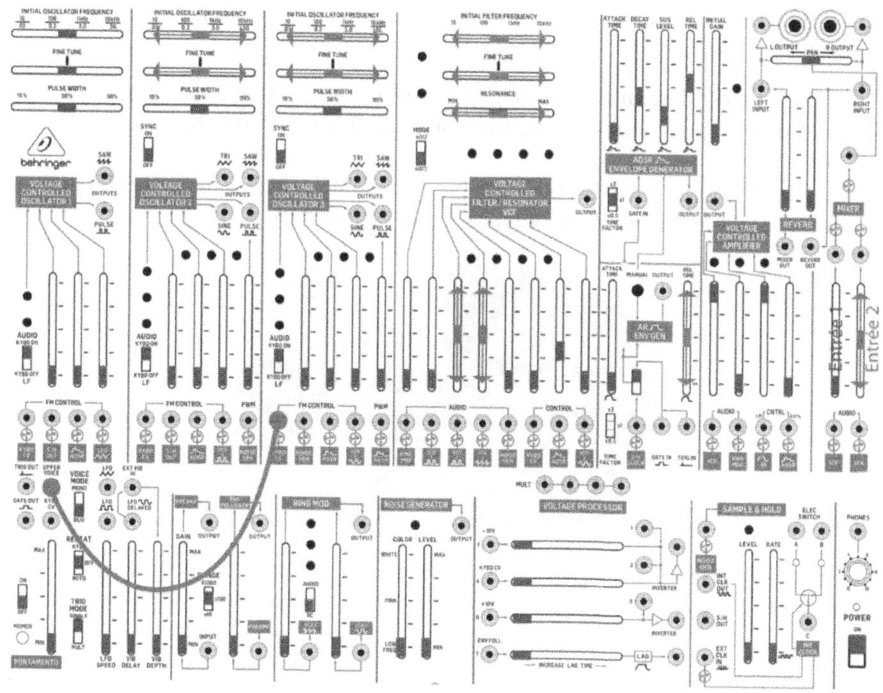

Figure 4.1. *Duophony with the Behringer 2600. For a color version of this figure, see www.iste.co.uk/reveillac/synthesizers2.zip*

Instructions and comments are as follows:

– place the faders and switches as shown in Figure 4.1;

– connect the UPPER VOICE output to the KYBD CV input of oscillator 3 using a patch cable;

– you can modify the balance between the sound of the two VCOs by altering the VCO1 and VCO2 faders;

– the frequencies can be different (tuning to the third, the fifth, etc.), although generally the VCOs can be in unison on the same octave or one or more octaves at a higher or lower pitch;

– the timbre of the sound can be modified by altering the frequency and resonance of the filter;

– the VCA release can also be adjusted by altering its fader;

– do not forget to adjust the VCA output level and set the VOICE MODE switch to the DUO position;

– if you hold a key in the lower part of your keyboard, you can play a melody simultaneously; in duophony mode, two notes can be played together.

NOTE.– On an original ARP 2600, a Korg or specific clones, the UPPER VOICE output is located to the left of the keyboard (see Figure 4.2).

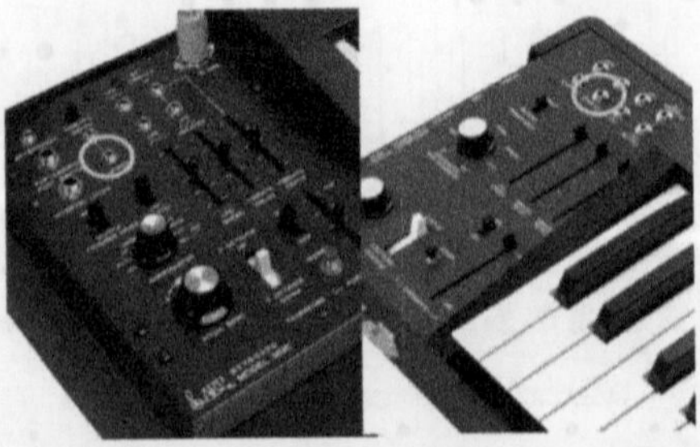

Figure 4.2. *UPPER VOICE output on two different ARP/Korg keyboards. For a color version of this figure, see www.iste.co.uk/reveillac/synthesizers2.zip*

4.1.2. *Novation Bass Station II*

The patch presented for this exercise uses the Bass Station II paraphony mode and is associated with an LFO.

As a reminder, this mode is available by pressing the FUNCTION key, then twice pressing on the keyboard key, D (D) sharp (Osc 1-2 Sync).

Then, choose the value P 1 on the LED display in the Master section and press the FUNCTION key to validate.

Figure 4.3. *The activated paraphony mode (P1) of the Bass Station II. For a color version of this figure, see www.iste.co.uk/reveillac/synthesizers2.zip*

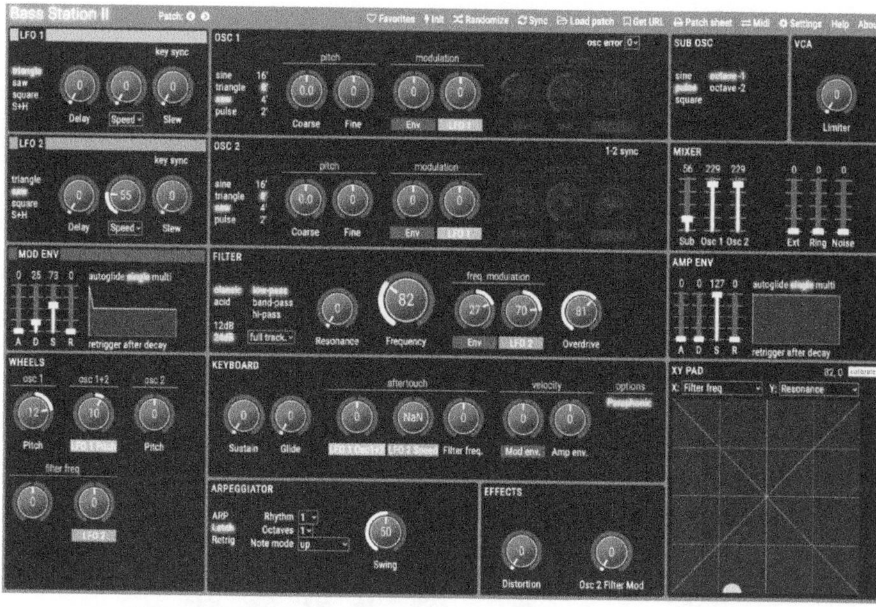

Figure 4.4. *Paraphony on the Novation Bass Station II. For a color version of this figure, see www.iste.co.uk/reveillac/synthesizers2.zip*

Instructions and comments are as follows:

– set the values of the different controls as shown below;

– LFO 2: SPEED = 60;

– OSC 1: SAW 8';

– OSC 2: SINE 8;

– SUB OSC: PULSE, OCTAVE-1;

– MOD ENV: D = 25; S = 73; SINGLE;

– AMP ENV: S = 127, SINGLE;

– FILTER: CLASSIC, LOW-PASS, 24 dB, FULL TRACK;

– FILTER: RESONANCE = 0, FREQUENCY = 82, FREQ. MODULATION – ENV = 27, LFO 2 = 70; OVERDRIVE = 81;

– MIXER: SUB = 56, OSC 1 = 229, OSC 2 = 229;

– ARPEGGIATOR: LATCH; RHYTHM = 1; OCTAVES = 1; NOTE MODE = UP; SWING = 50;

– all other parameters: 0 or default value;

– hold a key for bass and play a melody note for note. In paraphony mode, two notes can be played together;

– by modifying the volumes of each oscillator in the MIXER section, you can configure the general balance;

– also try modifying the range of one of the oscillators and its waveform;

– remember to set the output volume to an appropriate value.

4.1.3. *Behringer Neutron*

Below you will find two paraphony patches dedicated to the Behringer Neutron. They come, with some modifications, from the GitHub Patch Library[1].

4.1.3.1. *Patch No. 1*

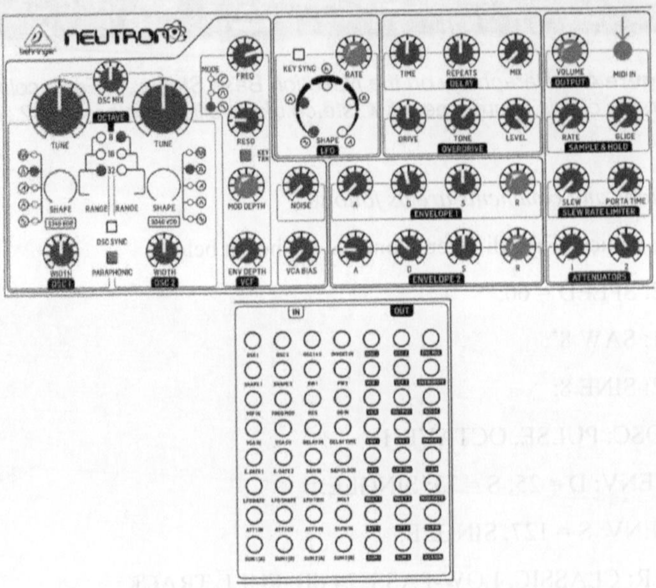

Figure 4.5. *Simple patch for the Behringer Neutron in paraphony mode. For a color version of this figure, see www.iste.co.uk/reveillac/synthesizers2.zip*

1 Available at: www.patch-library.net.

This paraphony patch on the Behringer Neutron uses the existing pre-wiring. No cable is required on the connection matrix; we owe this to Avrilcadabra, a contributor to the patch-library.net site.

Instructions and comments are as follows:

– set the controls as shown in Figure 4.5;

– you can modify the MOD DEPTH settings of the filter (VCF), the RATE of the low-frequency oscillator (LFO) and the two releases (R) of envelope generators 1 and 2 to obtain the sound that suits you.

4.1.3.2. *Patch No. 2*

This second patch is inspired by one created by Pieslice, another contributor to the patch-library.net site. Its sound is oriented more toward trance or ambient musical styles.

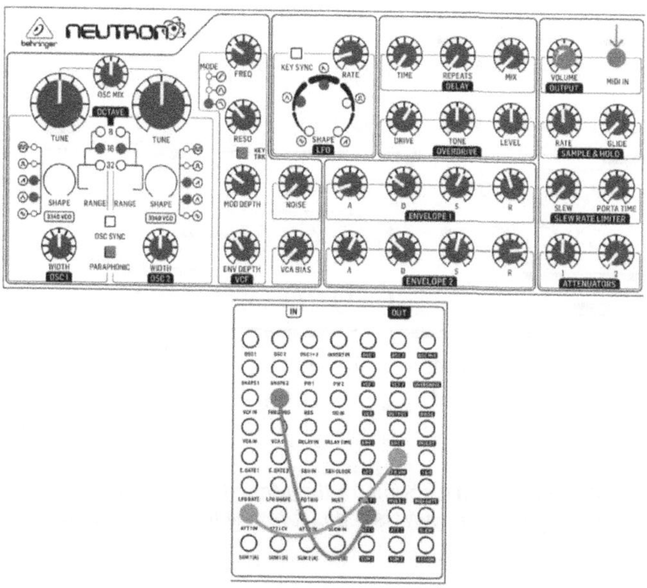

Figure 4.6. *Advanced paraphony mode for the Behringer Neutron. For a color version of this figure, see www.iste.co.uk/reveillac/synthesizers2.zip*

Instructions and comments are as follows:

– set the controls as shown in Figure 4.6;

– on the matrix, make the following LFO connections: UNI to ATT1 IN and ATT1 to FREQ MOD;

– in this patch, attenuator 1 alters the LFO, which conditions the filter cutoff frequency;

– hold down a key in the lower part of the keyboard, like a held note on a church organ, and play a melody in single notes;

– you can modify the setting of attenuator one and the rate of the LFO to obtain the appropriate sound reproduction.

4.1.4. *Arturia MatrixBrute*

The MatrixBrute has a paraphony mode that isolates the three oscillators, allowing up to three notes to be played simultaneously. Like all paraphony modes, the three oscillators share the same filters and VCA setting.

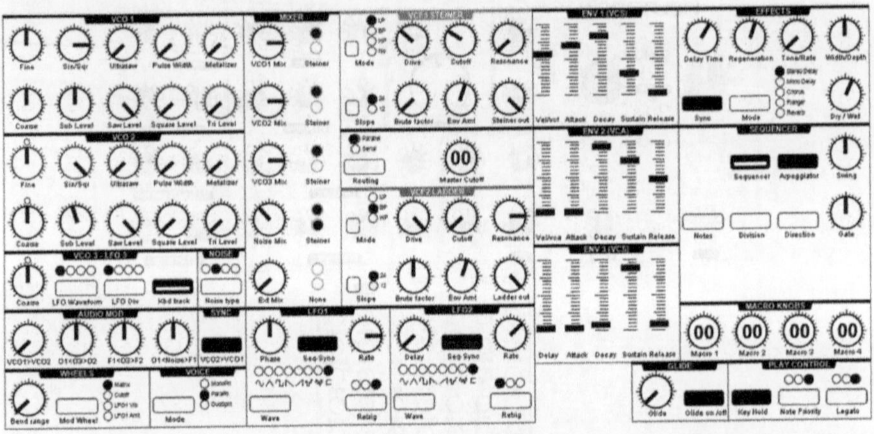

Figure 4.7. *Paraphonic sound for the MatrixBrute. For a color version of this figure, see www.iste.co.uk/reveillac/synthesizers2.zip*

Instructions and comments are as follows:

– amend the settings as shown in Figure 4.7;

– remember to activate keyboard tracking (KBD TRACK) and paraphony voice mode (VOICE PARAPH);

– do not forget to set the stereo delay effect (ANALOG EFFECTS – STEREO DELAY);

– all three oscillators use the STEINER filter;

– the noise generator provides pink noise and uses two filters, STEINER and LADDER;

– you will need to set up the matrix assignments as shown in Table 4.2, and associate them with the settings in Figure 4.7;

From	To	From	To
LFO1	VCO1 Pitch	**Velocity**	Steiner Cutoff
LFO1	VCO2 Pitch	**M1**	VCO1 Ultra
LFO1	13	**M1**	VCO2 Ultra
LFO1	14	**M1**	Steiner Cutoff
LFO2	14	**M1**	Ladder Cutoff
ModWheel	Steiner Cutoff	**Exp1/M3**	15
ModWheel	Ladder Cutoff	**Exp2/M4**	14
Keyb/Seq	Steiner Cutoff	**Exp2/M4**	16

Table 4.2. *MatrixBrute matrix assignments*

– assign the following to parameters 13-16: 13 – VCO1 Coarse; 14 – Fx Delay Time; 15 – VCO1->Noise->VCF1; 16 – Fx Width/Depth;

– the M1 macro-setting acts on the VCO1 and VCO2 (ultra) oscillators and the two filters;

– the M3 macro-setting alters the modulation with the noise oscillator of VCO1 and VCF1;

– the M4 macro-setting affects the delay time of the delay effect (FX Delay Time) and its width/depth;

– hold down one or two keys on the left (low) of the keyboard and play a melody on the right (high). The sound of this patch lends itself to slow music, even a held note whereby each note can develop its full extent.

4.2. Exercise 15 – polyphony

Duophony and paraphony remain the prerogative of monophonic synthesizers, often offering additional sound processing options.

Exercise 15 will deal with true polyphony, a characteristic that does not mean the number of notes played simultaneously corresponds to an equivalent number of voices.

A sound can consist of one or more voices combined. It all depends on the architecture of the machine, whether hardware or software. The number of voices for a sound can be up to four, which divides your polyphony accordingly. For example, for a 64-voice synthesizer using four voices per note, the polyphony will equal 16 simultaneous notes.

There are also floating polyphony machines whereby the number of available voices varies depending on the needs in terms of sound reproduction.

Polyphony should not be confused with multitimbrality, which is the option of simultaneously producing different sounds based on one or more voices.

4.2.1. *Max/MSP*

Creating a polyphonic instrument with Max/MSP is not complex; however, several solutions are available, requiring more powerful processors.

In the following, we will review some of the options this platform offers.

4.2.1.1. *Abstractions multiples*

The most straightforward technique for creating a polyphonic synthesizer is using an abstraction as many times as the number of voices requires.

This simple solution works perfectly but has a major drawback; it uses a lot of the resources of the microcomputer the program is running on. Indeed, each copy of the abstraction remains active, regardless of its use.

Microcomputers have indeed seen their computing power increase over time, and the limitations encountered previously are now much less significant.

NOTE.– Using the "DSP CPU Monitor" tool included as an extra with the latest versions of Max/MSP, the resource consumption of digital signal processing (DSP) can easily be checked.

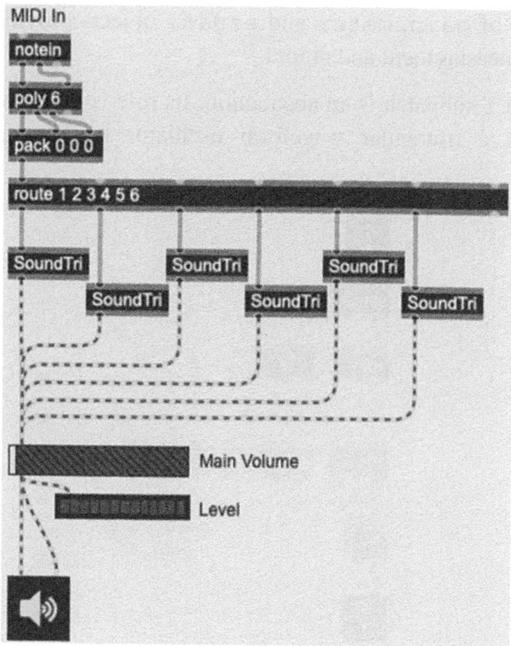

Figure 4.8. *A multi-abstraction polyphonic patcher*

Instructions and comments are as follows:

– start by building the main patcher;

– create the objects in Figure 4.8 and connect them as shown;

– the `notein` object captures the note from the MIDI keyboard;

– the `poly 6` object is connected to the two left outputs of the `notein` object (note and velocity). It allocates notes to different voices. Here, six voices were declared;

– the `pack` object is connected to the input of the `poly 6` object and the output of the `route 1 2 3 4 5 6` object. It builds a list containing the voice number, the note and its velocity;

– the `route 1 2 3 4 5 6` object orients the note and its velocity toward the chosen voice, from 1 to 6 depending on the number of keys pressed on the keyboard by the user;

– the subpatch composed of the `SoundTri` abstraction generates the note. It retrieves the note and velocity from the `route 1 2 3 4 5 6` objects. It is duplicated six times for each of the six voices;

– the sequence of `gain`, `meter` and `ezdac~` objects provide general volume adjustment, level measurement and output;

– the `SoundTri` subpatch is an abstraction. Its role is to convert the note into a frequency sent to a triangular waveform oscillator by associating it with an envelope;

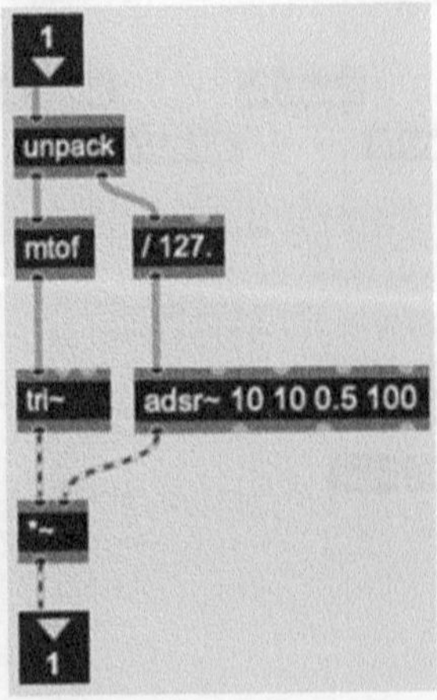

Figure 4.9. *The SoundTri abstraction*

– the construction of this subpatch is basic and resembles sequences of objects already encountered;

– the `inlet 1` object receives the note and velocity from the `route` object;

– the `unpack` object, following the `inlet 1` object, breaks down the list into a note and its velocity;

– the `mtof` object retrieves the note at the output of the `unpack` object and converts the note into frequency;

– the `tri~` oscillator generates a triangular waveform at the frequency determined by the `mtof` object;

– the `/ 127.` object, connected to the right output of the `unpack` object, converts velocity to a value between 0 and 1;

– the `adsr~ 10 10 0.5 100` object defines a default ADSR envelope where A = 10 ms, D = 10 ms, S = 0.5 (50%) and R = 100 ms;

– the following `*~` object associates the envelope with the triangular signal;

– be careful when adjusting the volume; the simultaneous use of several abstractions generates a high level, which can lead to the risk of saturation. You can control the latter with the `meter~` object;

– do not forget to test the polyphony by pressing several keys on the keyboard;

– you can check that no more notes (pressed keys) are considered beyond six.

4.2.1.2. *Using the poly~ object*

The `poly~` object creates several copies of the subpatch within the same object. It limits its activities to only the voices the user requests, which saves microcomputer processor resources.

In this patcher, more precise envelope control is possible, allowing the user to control the four envelope parameters: attack, decay, sustain and release (ADSR).

On the other hand, a drop-down menu allows the oscillator waveform to be chosen: sinusoidal, sawtooth, triangular or rectangular.

Instructions and comments are as follows:

– as usual, the `notein` object captures the note(s) emitted from the MIDI keyboard;

– a `number` object output from the `notein` object allows you to control the MIDI number of each note;

– a `pack` object creates a list combining note and velocity;

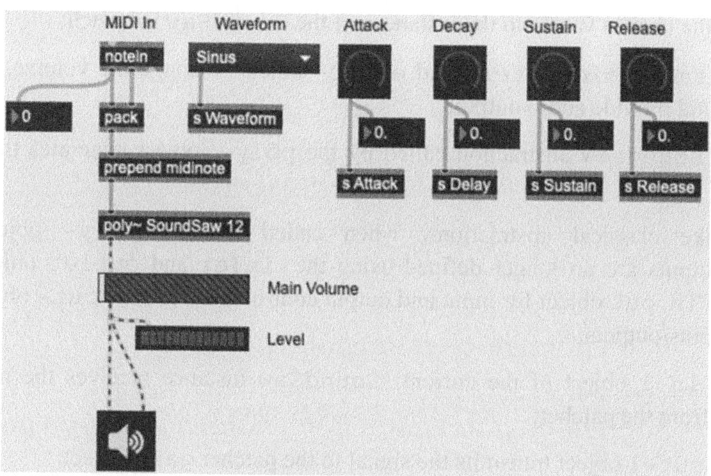

Figure 4.10. *A polyphonic patcher using the poly~ object*

– a prepend midinote object preloads the note and its velocity to transmit it to the following poly~ SoundSaw 12 object;

– the poly~ SoundSaw 12 object creates 12 instances (copies) of the SoundSaw subpatch, becoming the 12 voices available to the user. When a message is received from the previous prepend midinote object, it goes through the copies of the subpatch until it finds an available one (voice not busy processing a previous note) to transmit the preloaded note and velocity;

– the umenu Waveform object allows you to choose the waveform (none, sine, sawtooth, triangular, rectangular);

– the s Waveform object transmits the number of the chosen waveform (0 to 4) to the SoundSaw subpatch instance;

– the four dial objects control the four parameters (ADSR) of the envelope generator;

– the dial Attack object has a size of 2001 and a minimum value of 0;

– the dial Decay object has a size of 1001 and a minimum decimal value of 0;

– the dial Sustain object has a size of 1 and a minimum decimal value of 0;

– the dial Release object has a size of 2001 and a minimum value of 0;

– the four flonum objects following the dial objects display the values chosen by the user;

– the four s Attack, s Delay, s Sustain and s Release objects transmit the chosen values to the instances of the SoundSaw subpatch;

– the three gain, meter~ and ezdac~ objects manage the volume, display its level and provide audio output;

– the Soundsaw abstraction called by the poly~ object generates the audio signal;

– unlike classical abstractions, when called from a poly~ object, the inputs–outputs are no longer defined using the inlet and outlet objects but using the in, out object for input and output control and in~ and out~ objects for signal inputs/outputs;

– the in 1 object of the current SoundSaw instance receives the note and velocity from the patcher;

– the out~1 object transmits the signal to the patcher gain object;

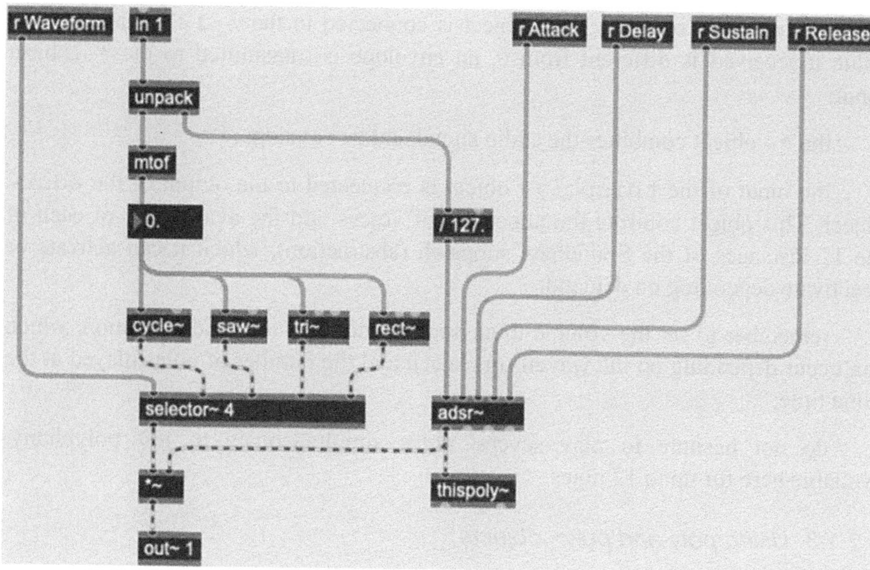

Figure 4.11. *The SoundSaw abstraction*

– the five `r Waveform`, `r Attack`, `r Decay`, `r Sustain` and `r Release` objects, respectively, receive data from the `s Waveform`, `s Attack`, `s Delay`, `s Sustain` and `s Release` objects from the patcher;

– the `unpack` object divides the note-velocity list into a MIDI note number transmitted to the `mtof` object and a velocity value transmitted to the `/ 127` object, which will be divided by 127 to obtain a value between 0 and 1;

– the `mtof` object receives the MIDI number of the note and transforms it into a frequency value displayed by the following `flonum` object;

– the four `cycle~`, `saw~`, `tri~` and `rect~` objects, connected to the output of the `flonum` object, are the four oscillators that provide the waveform chosen by the user from the patcher `umenu Waveform` object. The outputs of these four objects are connected to the right inputs of the `selector~ 4` object;

– the left input of the `selector~ 4` object receives the value (from 0 to 4) corresponding to the option chosen by the user in the `umenu Waveform` object of the patcher and selects the appropriate waveform, which arrives on one of its four right inputs;

– the four right inputs of the `adsr~` object are connected to the outputs of the four `r Attack`, `r Decay`, `r Sustain` and `r Release` objects to recover the transmitted values;

– the left input of the `adsr~` object is connected to the `/ 127` object. If the value it received is different from 0, an envelope is transmitted to the `*~` object input;

– the `*~` object combines the audio signal and its envelope;

– the input of the `thispoly~` object is connected to the output of the `adsr~` object. This object controls the allocation of voices and the availability of each of the 12 instances of the SoundSaw subpatch (abstraction), which it can activate or deactivate depending on demand;

– remember to set the volume to an appropriate level to avoid saturation, which can occur depending on the waveform chosen and the number of notes played at the same time;

– do not hesitate to play several notes simultaneously to test polyphony, available here for up to 12 notes.

4.2.1.3. *Using poly and poly~ objects*

This exercise uses most of the objects from the previous exercise but both `pack` and `prepend midinote` objects are replaced by three others: `poly 12`, `pack 0 0 0`, and `target $1, $2 $3` objects, as you can see in Figure 4.12.

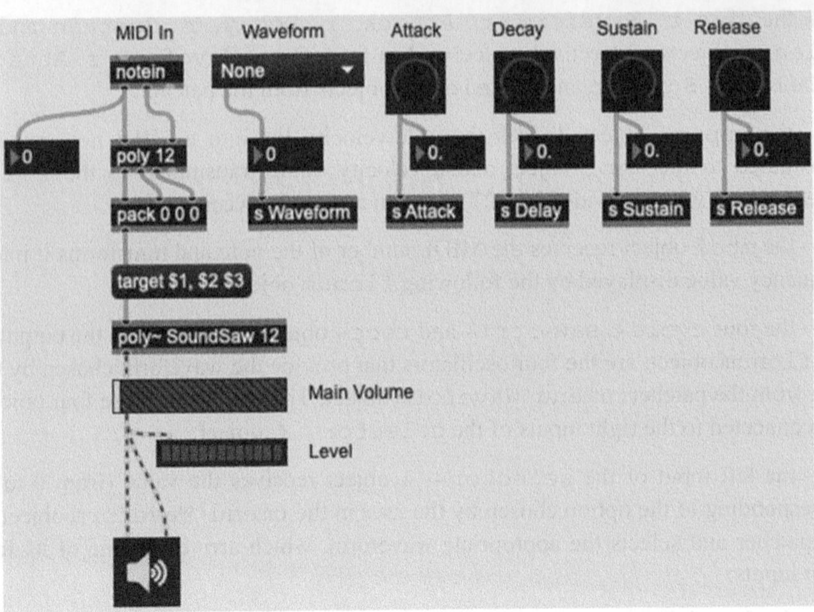

Figure 4.12. *A patcher using poly and poly~ objects*

Instantiation via the poly~ object and SoundSaw abstraction (see Figure 4.11) remain unchanged.

Instructions and comments are as follows:

– we will not return to the functions and links of the objects already presented in the previous exercise in Figures 4.10 and 4.11;

– the poly 12 object allocates the data received by the notein object, note and velocity with 12 individual voices;

– the pack 0 0 0 object creates a list consisting of the channel number, the note, and its velocity; by default, the list contains 0 0 0;

– the message target $1, $2 $3 object retrieves data from the pack object and creates a message with the target object attribute so that the poly~ SoundSaw 12 object can route the values to an available SoundSaw instance;

– all other objects of the patcher and SoundSaw subpatch have functions identical to those described in the previous exercise.

4.2.2. *Pure Data*

For this polyphony exercise with Pure Data, two solutions will be proposed: the first using a poly object and multiple abstractions, and the second using a poly object and a clone object dynamically creating instances independent of the abstraction.

4.2.2.1. *Using the poly object*

This patch uses the poly object and a note abstraction to generate the audio signal. For this exercise, polyphony is limited to six voices.

Instructions and comments are as follows:

– a notein object retrieves the MIDI number of each note played by the user;

– a number object, at the output of the notein object, displays the MIDI number of the note played;

– the output of the number object is connected to the input of the poly 6 1 object, which allocates a voice number to the different notes played and retrieves the MIDI number of each note and its velocity;

– the outputs of the poly 6 1 object are the voice number, note and velocity from left to right in that order;

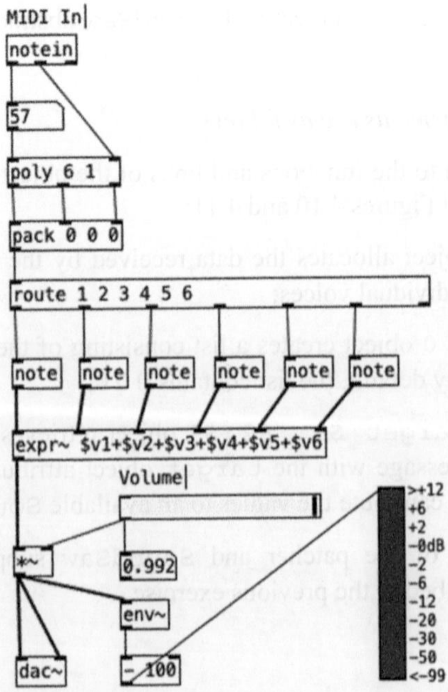

Figure 4.13. *A six-voice polyphonic patch*

– the `pack 0 0 0` object connected to the output of the `poly 6 1` object creates a list associating voice number, note and velocity. By default, these different parameters are null;

– the `route 1 2 3 4 5 6` object is connected to the output of the `pack` object, where it retrieves the list's contents, directing the note number and velocity to the different `note` abstractions. The voice number in the list allows the choice of one of the six abstractions available at its output. The rightmost output of the `route` object is the default output, but it is not used here;

– the `expr~ $v1+$v2+$v3+$v4+$v5+$v6` object evaluates and adds the different audio signals from the abstractions;

– the `*~`, `dac~`, `slider Volume`, `number`, `env~`, `- 100` and `meter~` objects control the volume of the audio signal, its level measurement and output. This set of objects has already been encountered in section 2.4.5.

NOTE.– The `poly` object has a voice-stealing parameter (reallocation of voices). If the second parameter of the `poly` object is 0, no reallocation is made beyond those planned (in our exercise, it is equal to six). If it is 1, a reallocation is made. To check this, hold down six keys on the keyboard, then press a seventh; when the reallocation is valid (1), a new sound is heard and the sound of the first key pressed disappears. If the reallocation is invalid (0), the seventh key pressed does not emit any sound, and there is no reallocation.

– the `note` abstraction is based on a classic construction that generates a note and an associated envelope;

– the `inlet` object receives note number and velocity data from the `route 1 2 3 4 5 6` object;

– the `unpack 0 0` object connected to the output of the `inlet` object disassembles the list by separating the note number (left output) from the velocity (right output);

– the `number` object displays the MIDI note number;

– the `mtof` object converts the MIDI note number to a frequency;

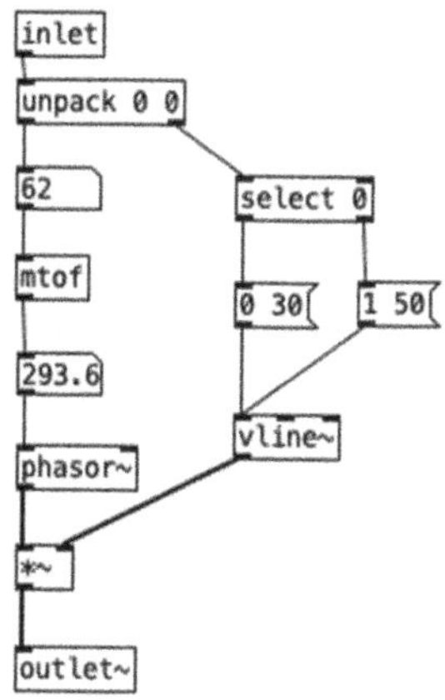

Figure 4.14. *The `note` abstraction of the polyphonic patch*

– the `number` object at the input of the `mtof` object displays the frequency;

– the `phasor~` object generates a sawtooth signal depending on the received frequency;

– the `select 0` object tests the velocity and selects the left or right output. If the velocity is 0, the left output is active; otherwise, the right output is selected;

– the `message 0 30` object sets a descending audio signal ramp of 30 ms duration;

– the `message 1 50` object defines an ascending audio signal ramp of 50 ms;

– the `vline~` object recovers the ramp message; either it cuts the signal according to a downward slope of 30 ms, or it increases the audio signal at an upward slope of 50 ms. This object acts as an envelope generator;

– the `*~` object combines (multiplies) the audio signal and its envelope;

– the `outlet~` object transmits the audio signal to the `expr~ $v1+$v2+$v3+$v4+$v5+$v6` patch object;

– be careful not to set the volume too high. When several notes are played simultaneously, the volume increases and can lead to saturation. The `meter~` is there to check the level.

4.2.2.2. *Using poly and clone objects*

Another way to create a polyphonic instrument with Pure Data is by using the `clone` object, which dynamically creates abstraction instances based on several requested voices.

For this exercise, the `PolySynth` abstraction will be used; it generates the waveform and defines its envelope.

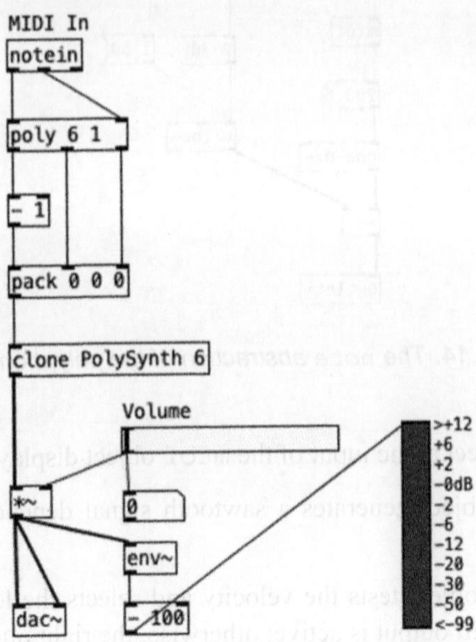

Figure 4.15. *A polyphonic patch using the* `poly` *and* `clone` *objects*

Instructions and comments are as follows:

– a notein object retrieves the MIDI number of each note played by the user;

– the output of the notein object is connected to the input of the poly 6 1 object, which allocates a voice number to the different notes played and retrieves the number of each note and its velocity;

– the outputs of the poly 6 1 object are the voice number, note and velocity from left to right in that order;

– the - 1 object, connected to the left output of the poly 6 1 object, retrieves the voice number, from which it subtracts 1 in order to manage allocations between 0 and 5 (6 voices);

– the pack 0 0 0 object, connected to the output of the poly 6 1 object, creates a list associating voice number, note and velocity. By default, these different parameters are 0;

– the PolySynth clone 6 controls the six polyphony voices dynamically, ensuring instantiation and their allocations or reallocations;

– the *~, dac~, slider Volume, number, env~, - 100 and meter~ objects manage the volume of the audio signal, its level measurement and output. This set of objects has already been encountered in the previous exercise and section 2.4.5;

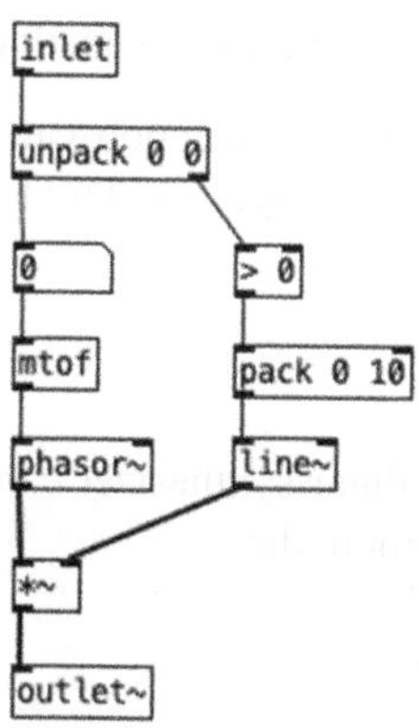

Figure 4.16. *The* PolySynth *abstraction*

– the PolySynth abstraction is based on a classic construction that generates a note and an associated envelope;

– the `inlet` object receives note number and velocity data from the `route 1 2 3 4 5 6` object;

– the `unpack 0 0` object, connected to the output of the `inlet` object, disassembles the list by separating the note number (left output) from the velocity (right output);

– the `number` object displays the MIDI note number;

– the `mtof` object converts the MIDI note number to a frequency;

– the `number` object, placed at the output of the `mtof` object, displays the frequency;

– the `phasor~` object generates a sawtooth signal, depending on the received frequency;

– the `> 0` object verifies if the velocity is greater than 0. In this case, a list is created by the `pack 0 10` object;

– the input of the `pack 0 10` object is connected to the output of the `> 0` object;

– the `line~` object retrieves at its input the list generated by the `pack 0 10` object and generates an ascending ramp (envelope), which the audio signal will use;

– the `*~` object combines (multiplies) the audio signal and its envelope;

– the `outlet~` object transmits the audio signal to the `*~` patch object;

– adjust the volume with caution; saturation can occur when several notes are played simultaneously. The `meter~` object is there to check the level.

4.2.3. *VCV Rack*

The VCV Rack software modular synthesizer can work in polyphonic mode for up to 16 voices. By default, it is monodic.

4.2.3.1. *Simple polyphonic rack*

The rack in Figure 4.17 shows a classic synthesizer made up of an oscillator (VCO), a filter (VCF), two envelope generators (ADSR) and an amplifier (VCA). At its input, a module manages a MIDI keyboard (MIDI-CV); at its output, it ensures audio processing (AUDIO).

Figure 4.17. *A polyphonic VCV rack. For a color version of this figure, see www.iste.co.uk/reveillac/synthesizers2.zip*

Instructions and comments are as follows:

– create a new rack and set up the different modules as shown in Figure 4.17;

– with cables, create the connections shown in Table 4.3;

– the envelope generators are differentiated according to their destination – VCF or VCA – on the left and the right in Figure 4.17;

From	To	From	To
MIDI-CV–GATE	ADSR (VCF)–GATE	**ADSR (VCA)–ENV**	VCA–Control
MIDI-CV–GATE	ADSR (VCA)–GATE	**VCO–SAW**	VCF–IN
MIDI-CV–V/OCT	VCO–V/OCT	**VCF–LPF**	VCA–IN
ADSR (VCF)–ENV	VCF–CUT	**VCA–OUT**	AUDIO–L/MON

Table 4.3. *Connections to be made in the rack*

– to activate polyphony, for example, with four voices, right-click on the MIDI-CV module and, in the menu, choose Polyphony channels, then 4 (see Figure 4.18);

– the color of the cables should become more intense, meaning that there are multiple connections;

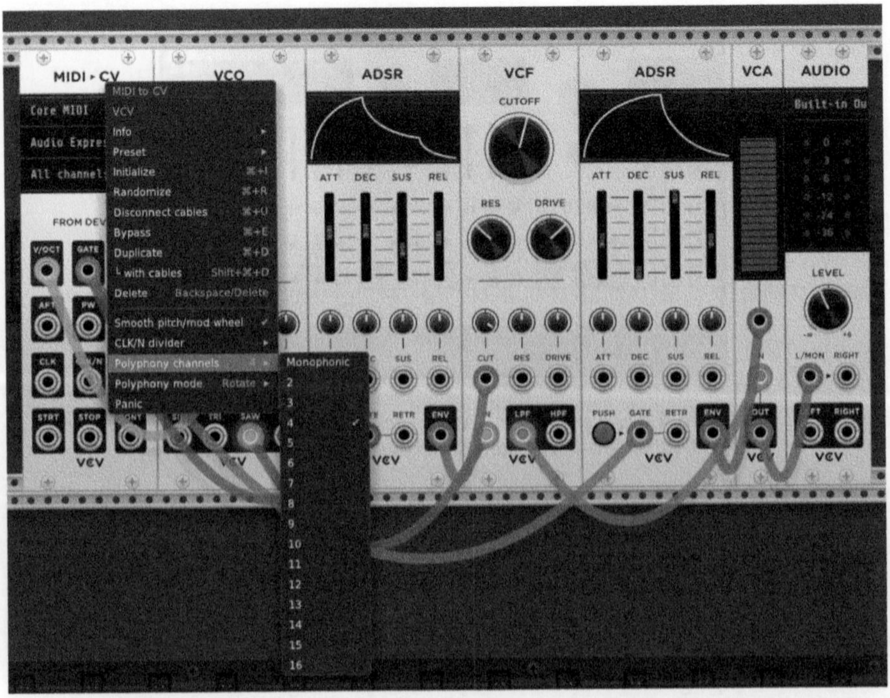

Figure 4.18. *Switching to four-voice polyphonic mode. For a color version of this figure, see www.iste.co.uk/reveillac/synthesizers2.zip*

– if you hover over each input–output point with a connection cable with your mouse cursor, you should see a message appear indicating the number of volts for each voice, as shown in Figure 4.19;

– try playing several notes simultaneously to test polyphony. You can increase or decrease the number of voices; however, the higher the number of voices, the more demand is placed on your computer memory and processors;

– you will notice that the VCA module display multiplies according to the number of voices defined, showing you the allocations of each of them when pressing each key on the MIDI keyboard;

– as we have already seen, you can choose a different waveform for the VCO output. Remember to adjust the parameters of the envelope generators (ADSR), those of the filter (VCF) and the volume of the VCA.

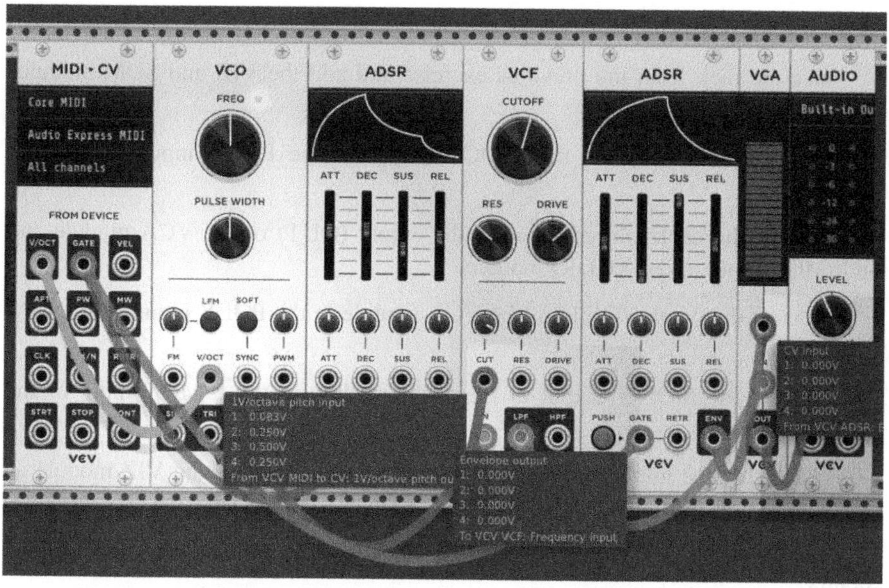

Figure 4.19. *Checking the presence of the four available voices on the inputs–outputs of the different modules. For a color version of this figure, see www.iste.co.uk/reveillac/synthesizers2.zip*

4.2.3.2. *Polyphonic rack with visualization and summator*

To best visualize the polyphonic mode, VCV Rack has two specific modules: the VIZ module, which shows the number of voices available, and the SUM module, which sums the different voices while managing the overall level. These two modules are part of the VCV Rack Fundamental Blocks library.

Figure 4.20. *Polyphonic rack with VIZ and SUM modules. For a color version of this figure, see www.iste.co.uk/reveillac/synthesizers2.zip*

Instructions and comments are as follows:

– take the rack from the previous exercise and add the VIZ and SUM modules (Fundamental Blocks Library);

– using a cable, connect the V/OCT output to the POLY input of the VIZ module;

– remove the cable placed between the output (OUT) of the VCA module and the L/MON input of the AUDIO module;

– connect the output (OUT) of the VCA module to the POLY input of the SUM module;

– connect the MONO output of the SUM module to the L/MON input of the AUDIO module;

– check that the number of voices is correctly displayed on the VIZ module as well as on the SUM module (CHAN);

– adapt the overall polyphony level with the LEVEL button on the SUM module.

4.2.4. *Reaktor*

With Reaktor, polyphony can be obtained by multiplying the sound generating circuits, for example, four circuits made up of 1 × VCO, 1 × VCF, 1 × VCA and 1 × GE, to which we will add a mixer, a dynamic voice allocation module, a level control module and a reverb to give a little depth to the sound reproduction.

The V Split module allows you to dynamically allocate five voices to automatically reallocate a voice when it is made available again by releasing a key on the MIDI keyboard. It was created by David Forrester and is based on MIDI management, called Note Allocator, developed by Salamanderanagram. This module is available from the Native Instruments user library[2].

The reverb module is also available from the library; this is the Pimped Rounds Reverb module, developed by Thala Estra, which we have already seen in section 3.3.1.6.

2 Available at: www.native-instruments.com/en/reaktor-community/reaktor-user-library/.

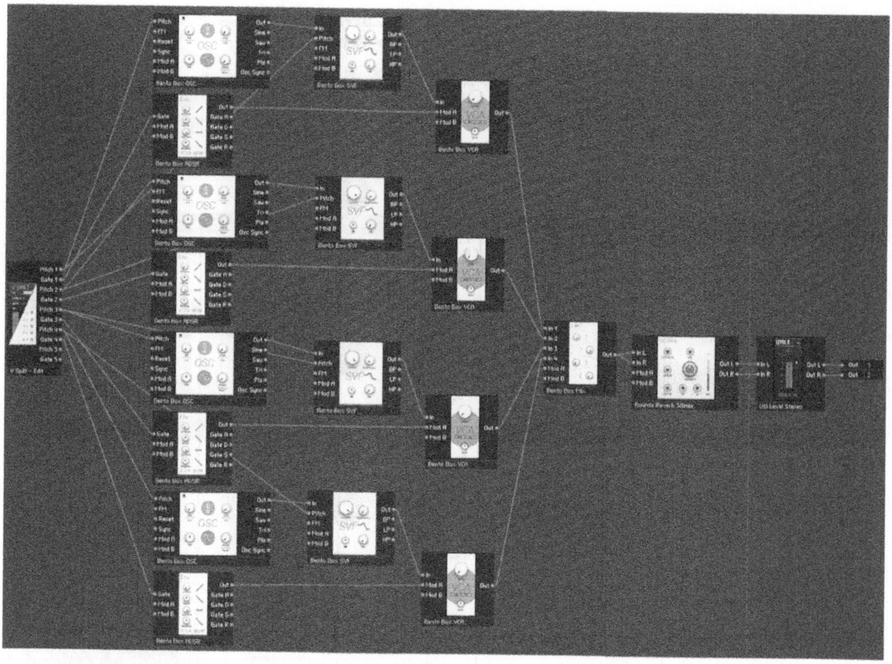

Figure 4.21. *A four-voice polyphonic Reaktor Blocks rack. For a color version of this figure, see www.iste.co.uk/reveillac/synthesizers2.zip*

Instructions and comments are as follows:

– all VCO, VCF, GE and VCA blocks and the mixer are from the Bento Box library (OSC, SVF, ENV, VCA and MIX);

– four subsets of blocks OSC, SVF, ENV and VCA manage each of the four voices;

– the V-SPLIT block is the input block and captures data from the MIDI keyboard;

– the PITCH outputs 1, 2, 3 and 4 are connected to the PITCH inputs of the OSC and SVF blocks of each subset;

– the GATE 1, 2, 3 and 4 outputs are connected one by one to the GATE inputs of each ENV block;

– the outputs of each OSC block are connected one by one to the IN inputs of the SVF blocks;

– the OUT outputs of each SVF block are connected one by one to each of the IN inputs of the VCA blocks;

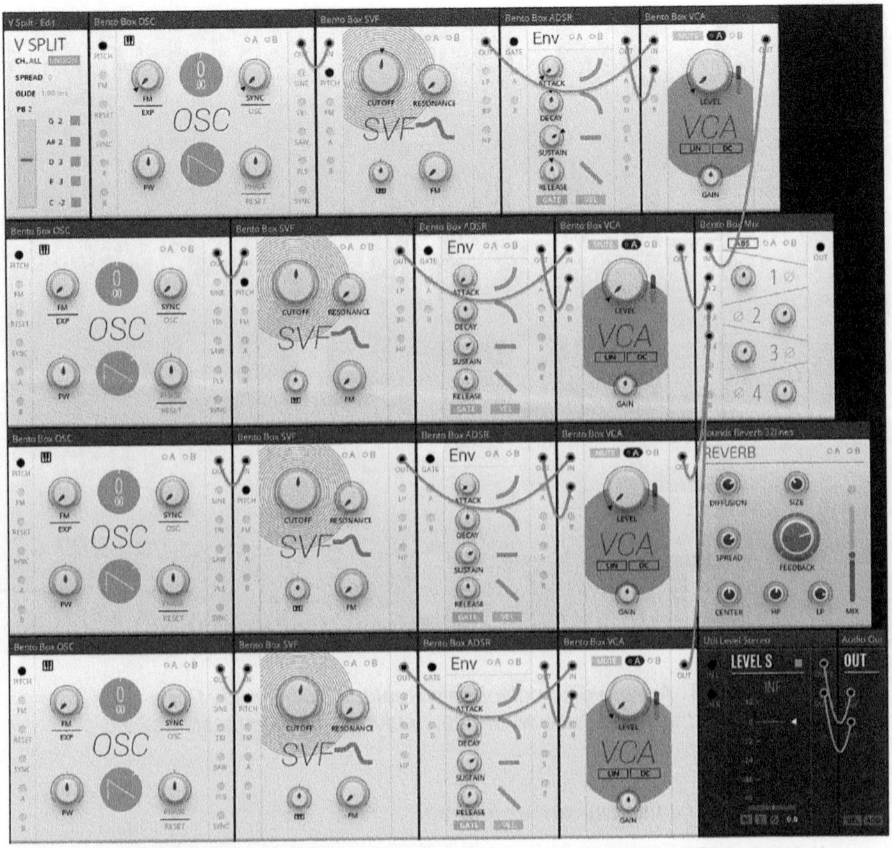

Figure 4.22. *Four-voice polyphonic rack in Panel mode. For a color version of this figure, see www.iste.co.uk/reveillac/synthesizers2.zip*

– the OUT outputs of each ENV block are connected one by one to each of the MOD A inputs of the VCA blocks;

– the OUT outputs of each VCA block are connected to the IN inputs 1, 2, 3 and 4 of the MIX block;

– the OUT output of the MIX block is connected to the two inputs IN L and IN R of the ROUND REVERB block;

– the OUT L and OUT R outputs of the ROUND REVERB block are connected to the IN L and IN R inputs of the LEVELS block;

– the OUT L and OUT R outputs of the LEVELS block are connected respectively to the OUT 1 and OUT 2 outputs;

– make sure to configure the blocks of each subset with the same parameters so there is no difference in tone between several notes;

– remember to place the fader of each modulation bus A of the VCA blocks in the high position (see Figure 4.22 and Figure 2.25 of Chapter 2) to have an operational envelope;

– adjust the different parameters of the reverb block;

– make sure the level on the LEVEL block is correctly adjusted.

4.3. Conclusion

Duophony and paraphony are interesting modes but they often remain incidental. They are only used to create specific effects or respond to occasional situations that require unexpected musical aspects. They are the prerogative of monodic synthesizers, in which optional functionality is added.

The polyphonic mode takes on a whole new dimension, bringing the musician towards new horizons, the possibility of playing a melody and its accompaniment, a bass line and its rhythm, and a piece with several voices together, etc.

As most musical instruments are monodic, such as woodwinds or brass, the piano or the organ, polyphonic by nature, present much superior dispositions in terms of musical composition.

Over time, from the 1970s to the present day, synthesizer designers have often sought to build polyphonic machines, although monophonic machines are still very present on the market.

A monodic synthesizer is more oriented toward the search for new sounds or new timbres to enrich a musical composition, unlike polyphonic synthesizers, which will easily succeed in replacing a piano, a harpsichord, a set of brass instruments, strings or even percussion.

You will no doubt have understood that the concept and use of a monodic synthesizer differ significantly from that of a polyphonic synthesizer, even if both are based on standard architectures and technical foundations.

5

Sequencers and Arpeggiators

A sequencer is a tool that can record and play a musical sequence. Although relatively similar, it is often confused with the arpeggiator, a tool that allows you to perform an arpeggio from a chord placed on a keyboard. Once the chord keys are released, a lock function allows the chord notes to loop until the next chord is played by the player.

Figure 5.1. *ARP sequencer (1976)*

In fact, sequencers are most often capable of integrating the functions of an arpeggiator. The arpeggiator often has advanced note scrolling functions, up, down, up then down, following a predefined pattern, etc.

In the 1970s, shortly after the first synthesizers were created, the first sequencers, limited to only a few notes (generally 8, 16 or 24) appeared, such as the ARP Sequencer (16 notes), the Korg SQ- 10 (24 notes), or the Moog 960 (24 notes). Very

For a color version of all the figures in this chapter, see www.iste.co.uk/reveillac/synthesizers2.zip.

quickly, more efficient models appeared, capable of stringing together several hundred notes, for example, the EMS Synthi AKS sequencer (256 notes).

These sequencers are all analog; some are still in use today, like the Korg SQ-1 or the A-155 module from Doepfer.

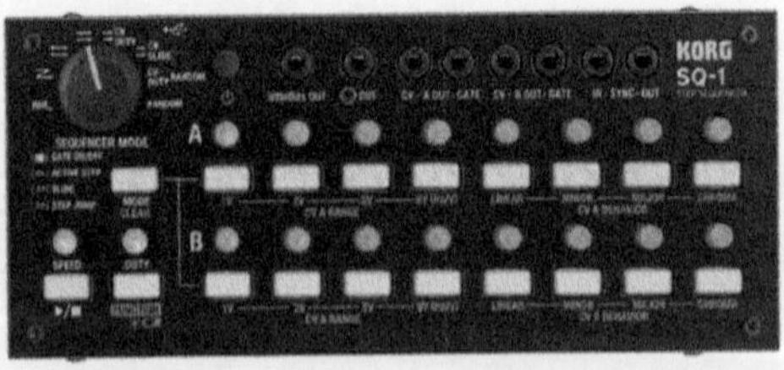

Figure 5.2. *Korg SQ-1 sequencer*

With the arrival of digital technology and the MIDI standard in the 1980s, the number of notes became almost unlimited while offering control of new parameters such as velocity, intensity, and after-touch.

MIDI sequencers then made their debut (Roland CSQ-100, Yamaha QY-10, Alesis MMT-8, etc.), and they were soon followed by software sequencers running on microcomputers.

Figure 5.3. *Yamaha QY10 (1990)*

Modern hardware synthesizers generally include a sequencer/arpeggiator. Some examples include the Novation Bass Station II, Arturia MatrixBrute, Moog Grandmother, etc.

5.1. Exercise 16 – sequencers and arpeggiators

This exercise will address sequencers/arpeggiators within the virtual modular synthesizers VCV Rack and Native Instruments Reaktor Blocks, then with the visual programming languages Max/MSP and Pure Data.

Several examples will be presented, showing various advanced functionalities with different implementations and configurations (absence/presence of a keyboard, pattern storage, arpeggios, presets, etc.).

5.1.1. *VCV Rack*

With the VCV Rack, many sequencer and arpeggiator modules are available; we will see the use of several of them. They can be complex and require the presence of a clock.

5.1.1.1. *Integrated clock sequencer*

For this first example, a simple sequencer module is chosen; it embeds its own clock and can generate three sequences of eight notes. This module is called SEQ3 and is one of the basic modules of the VCV Rack.

You will notice that, in this rack, the presence of a keyboard is not necessary. The sequencer plays the role of an instrumentalist, respecting the sequence composed by the user.

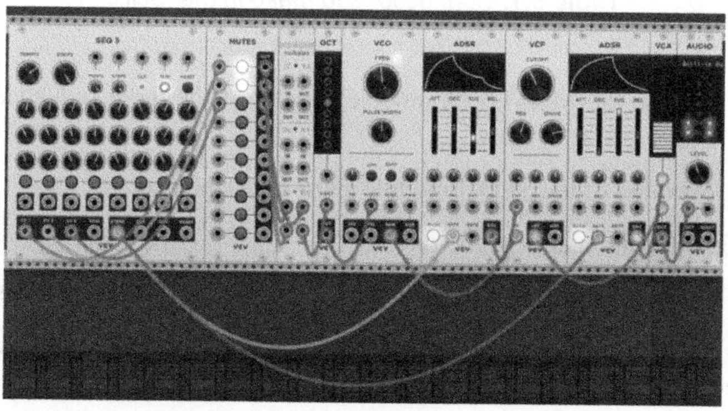

Figure 5.4. *VCV Rack with the SQ3 sequencer*

Instructions and comments are as follows:

– create a new rack;

– add the following modules: SEQ 3, MUTES, MULTIPLES, OCT, VCO, ADSR (× 2), VCF, VCA, and AUDIO. They all come from the default VCV Rack library;

– using cables, make the connections shown in Table 5.1;

– the envelope generators are differentiated according to their destination – VCF or VCA – on the left and right in Figure 5.4;

From	To	From	To
SEQ 3–CV1	MUTES–IN 1	**MULTIPLES–OUT**	OCT–V/OCT
SEQ 3–CV2	MUTES–IN 2	**OCT–OUT**	VCO–V/OCT
SEQ 3–CV3	MUTES–IN 3	**VCO–SAW**	VCF–IN
SEQ 3–STEPS	ADSR (VCF)–GATE	**ADSR (VCF)–ENV**	VCF–CUT
SEQ 3–STEPS	ADSR (VCA)–GATE	**VCF–LPF**	VCA–IN
MUTES–OUT 1	MULTIPLES IN 3: 1–IN 1	**ADSR (VCA)–ENV**	VCA–Control
MUTES–OUT 2	MULTIPLES IN 3: 1–IN 2	**VCA–OUT**	AUDIO–L/MON
MUTES–OUT 3	MULTIPLES IN 3: 1–IN 3	–	–

Table 5.1. *Connections to make in the rack*

– the buttons on the MUTES module select one of the three sequences of eight notes;

– the RUN button of the SEQ 3 module starts the sequencer;

– the TEMPO button adjusts the rate;

– the STEPS button defines the number of steps for the three sequences;

– the 24 knobs define the voltage and, therefore, the note (V/Oct) of each step in the sequence;

– the push buttons placed under each of the eight notes validate or invalidate their presence in the sequence;

– by right-clicking on the SEQ 3 module, you can control specific parameters, bypass, pre-selections, scrolling direction, etc.;

– be sure to set different voltages otherwise you will get a continuous sound by default;

– as we have already seen, you can also choose a different waveform for the VCO output. Remember to adjust the parameters of the octave changer (OCT), envelope generators (ADSR), those of the filter (VCF) and the VCA volume.

5.1.1.2. *External clock sequencer*

For this rack, we will use the Hampton Harmonics Progress module developed by Scott Hampton under an MIT License.

It is an eight-step monodic or polyphonic four-voice sequencer, allowing you to work with sequences of notes or chords.

This module requires an external clock, which an LFO or a dedicated module can simulate.

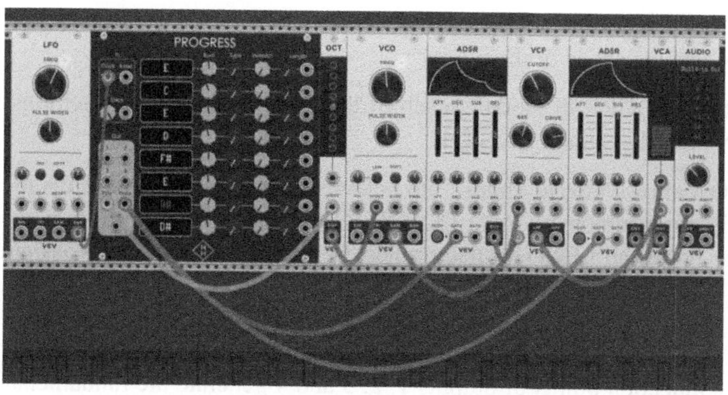

Figure 5.5. *VCV Rack with the Hampton Harmonics Progress sequencer*

Instructions and comments are as follows:

– create a new rack, or use the rack from the previous example and remove the SEQ 3, MUTES and MULTIPLES modules;

– here is a list of modules that must be present: LFO, PROGRESS, OCT, VCO, ADSR (2 ×), VCF, VCA and AUDIO. They are part of the default VCV Rack library, except PROGRESS, which you will have to download from the VCV Rack library;

– you will find all the connections to make in Table 5.2;

– the envelope generators are defined according to their destination – VCF or VCA – on the left and right in Figure 5.5;

From	To	From	To
LFO–SQR	PROGRESS–CLOCK	**VCO–SAW**	VCF–IN
PROGRESS–POLY	OCT–V/OCT	**ADSR (VCF)–ENV**	VCF–CUT
PROGRESS–GATE	ADSR (VCF)–GATE	**VCF–LPF**	VCA–IN
PROGRESS–GATE	ADSR (VCA)–GATE	**ADSR (VCA)–ENV**	VCA–Control
OCT–OUT	VCO–V/OCT	**VCA–OUT**	AUDIO–L/MON

Table 5.2. *Required connections*

– the LFO with its square signal serves as a clock for the sequencer; its frequency sets the speed of the notes;

– on the PROGRESS module, each of the eight notes has four settings: ROOT, TYPE, INVERSION and LENGTH;

– ROOT allows you to choose the octave and the base note, TYPE the type of the chord, major/minor 7th, etc., INVERSION defines the note to be played as the chord bass note, and LENGTH determines the number of clock beats assigned to the note for this step in the sequence;

– the STEP button defines the number of steps in the sequence;

– the connection points of outputs 1, 2, 3 and 4 generate the fundamental note of the chord, the second note, the third note and then the fourth, which is the same as the fundamental note for a 3-note chord. Inversion is not considered;

– you can also choose a different waveform for the VCO output. Remember to adjust the parameters of the octave changer (OCT), envelope generators (ADSR), those of the filter (VCF) and the VCA volume;

– instead of an LFO, you can choose a module from the VCV Rack library dedicated to managing a clock signal, such as LILT, CLKD or METRONOME.

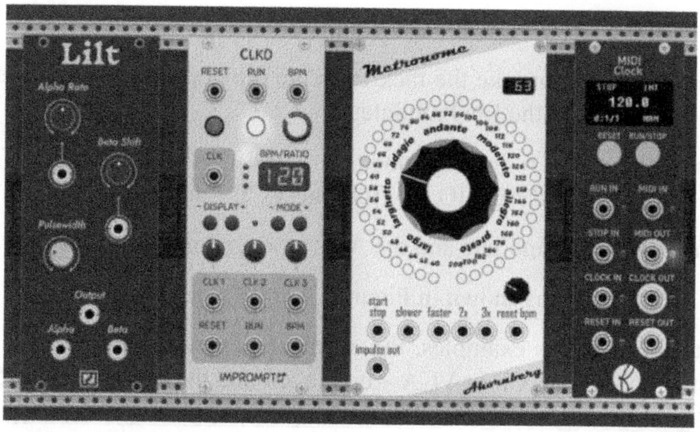

Figure 5.6. *Four clock signal generators (LILT by Fehler Fabrik, CLKD by Impromptu, METRONOME by Ahornberg and MIDI CLOCK by Kilpatrick-Toolbox)*

5.1.1.3. *Arpeggiator and external clock*

In this new rack, we will integrate an arpeggiator to capture a chord played on the keyboard and repeat it following a defined arpeggio. This agreement will have the same number of notes as channels assigned to the MIDI-CV module, from 2 to 16.

The chosen arpeggiator is the ARP module by Hampton Harmonics, developed by Scott Hampton, under an MIT license.

The LILT module by Fehler Fabrik, developed by Ross Cameron under a GPL 3.0 license, provides the clock signal necessary for the arpeggiator to function.

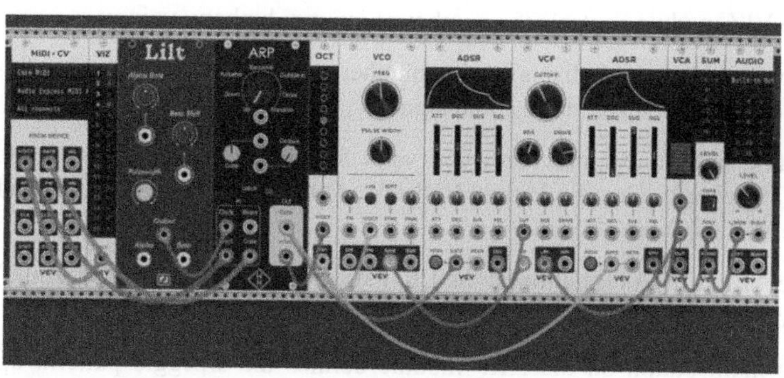

Figure 5.7. *A VCV rack using the ARP arpeggiator by Hampton Harmonics*

Instructions and comments are as follows:

– create a new rack, or take the rack from the previous example and remove the PROGRESS module and the LFO module;

– you must have the following modules: MIDI-CV, OCT, VCO, ADSR (× 2) VCF VCA SUM, VIZ and AUDIO, all from the VCV Rack default library;

– add the LILT clock module and the ARP arpeggiator module;

– make the connections shown in Figure 5.7 and Table 5.3;

– as with the previous examples, the envelope generators are defined according to their destination – the VCF or VCA – on the left and the right in Figure 5.7;

From	To	From	To
MIDI–CV–V/OCT	VIZ–POLY	**OCT–OUT**	VCO–V/OCT
MIDI–CV–V/OCT	ARP–V/OCT	**VCO–SAW**	VCF–IN
MIDI–CV–GATE	ARP–GATE	**ADSR (VCF)–ENV**	VCF–CUT
LILT–OUTPUT	ARP–CLOCK	**VCF–LPF**	VCA–IN
ARP–GATE	ADSR (VCF)–GATE	**ADSR (VCA)–ENV**	VCA–Control
ARP–GATE	ADSR (VCA)–GATE	**VCA–OUT**	SUM–POLY
ARP–V/OCT	OCT–V/OCT	**SUM–MONO**	AUDIO–L/MON

Table 5.3. *New connections to make on the VCV Rack*

– the VIZ module controls the number of input polyphony voices;

– the SUM module displays the number of output voices and adjusts the monophonic output level of the audio signal;

– define the number of voices you want for polyphony (maximum number of notes of the chords played) via the MIDI-CV module (see section 4.2.3.1) and check it on the VIZ module;

– the ALPHA RATE button of the LILT module adjusts the speed (tempo) of the arpeggio;

– the main button on the ARP module defines the arpeggio pattern: up (UP); down (DOWN); from bottom to top then from top to bottom (INCLUSIVE); bottom to top then top to bottom with the top note played only once (EXCLUSIVE); the lowest note, then the highest, then the second lowest and the second highest, and so on (OUTSIDE IN); in the order in which the notes arrive (ORDER); randomly (RANDOM);

– the OCTAVE button on the ARP module defines the number of octaves (1 to 4) over which the arpeggio will extend;

– the GATE button on the ARP module sets the length of the toggle signal from 0 (very short) to 1 (very long);

– when the LATCH switch on the ARP module is activated (high position), it allows you to hold the arpeggio of the last chord played;

– you will notice that the ARP module captures a polyphonic signal (chord) to output a monodic signal (succession of notes);

– you can also choose a different waveform for the VCO output. Remember to adjust the parameters of the octave changer (OCT), envelope generators (ADSR), filter parameters (VCF) and VCA volume.

5.1.2. *Reaktor*

For this exercise, three Reaktor racks will be used: the first will use the 8 STEPS sequencer from the Bento Box library, the second will integrate the 8 STEPS sequencer and the 4 Mods modulation sequencer, and the third will incorporate an arpeggiator.

5.1.2.1. *Eight-step sequencer*

This rack is configured with classic modules such as a VCO, a VCF, an envelope generator and a VCA.

No MIDI note capture module is present. The sequencer module is present to provide the melody notes, clocked via a clock module.

Instructions and comments are as follows:

– create a new rack;

– all the blocks VCO, VCF, GE, VCA and the mixer block come from the Bento Box library (OSC, SVF, ENV, VCA);

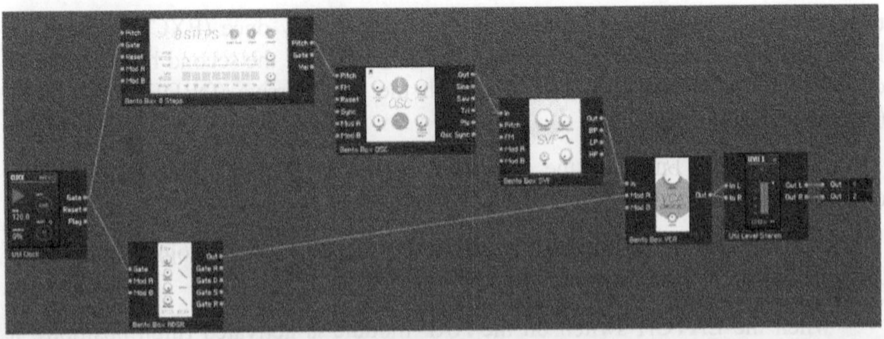

Figure 5.8. *Rack with a sequencer*

– the 8 STEPS sequencer block also belongs to the Bento Box library; it is called SEQ-BENTO 8 STEPS SEQUENCER;

– the CLOCK and LEVELS blocks are typically already present when creating a rack (UTIL-UTIL CLOCK, MIX-UTIL LEVELSTEREO);

– connect the GATE output of the CLOCK block to the GATE input of the 8 STEPS and ENV blocks;

– connect the output of the 8 STEPS block to the PITCH input of the OSC block;

– connect the OUT output of the OSC block to the IN input of the SVF block;

– connect the OUT output of the SVF block to the IN input of the VCA block;

– connect the OUT output of the ENV block to the MOD A input of the VCA block;

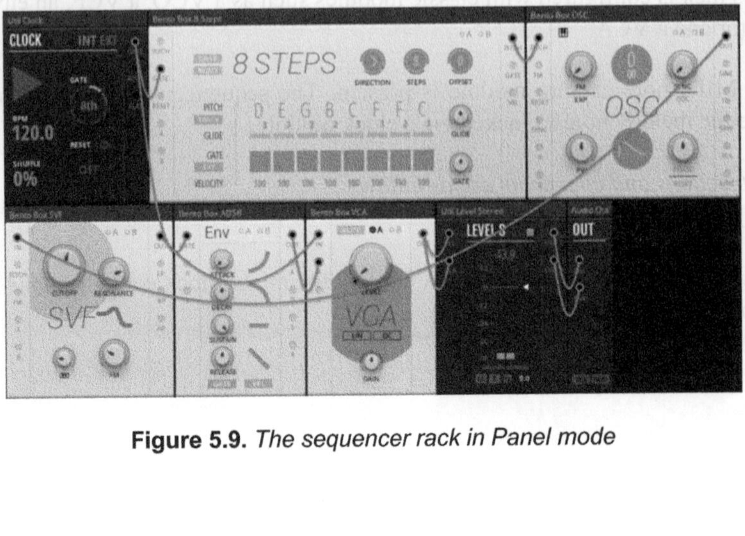

Figure 5.9. *The sequencer rack in Panel mode*

– connect the OUT output of the VCA block to the IN L and IN R inputs of the LEVELS block;

– connect the OUT L and OUT R outputs of the LEVELS block to the two OUT connection points of the OUT block;

– set a quantization from 4th to 64th on the CLOCK block with the GATE knob. You can also set a tempo in the BPM field and a shuffle percentage for a little more energy;

– by clicking the triangle-shaped button, you start the clock and the sequence. Make sure the clock is set correctly on INT (top and right of the module);

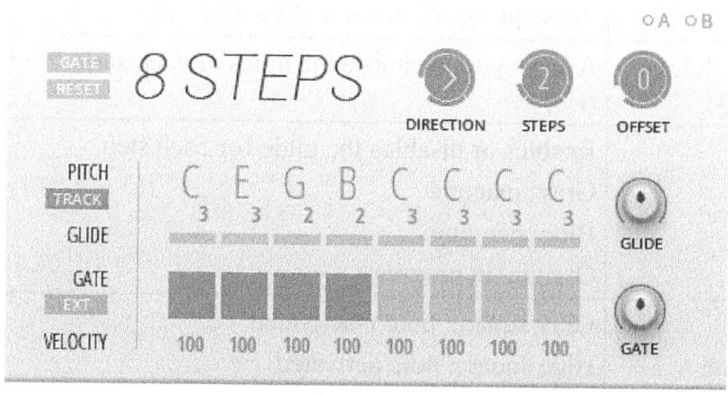

Figure 5.10. *The 8 STEPS module*

– regarding the 8 STEPS sequencer module (Figure 5.10), you will find information about the parameters of interest for this exercise in Table 5.4. To learn more about other parameters, see the Native Instruments Reaktor Blocks documentation;

– remember to choose different notes on the sequencer module otherwise you will obtain a series of continuous sounds;

– do not forget to configure the VCA modulation channel A (cursor up) to take into account the envelope parameters (ENV block);

– try different sequencer settings and change clock settings to get a different sequence flow.

NOTE.– You can connect the PITCH output of a NOTE IN block to the PITCH input of the 8 STEPS block and check what happens. Your sequencer can, in a way, become an arpeggiator to which you assign the first note of the chord. Indeed, if you record a four-note chord, for example, C, E, G, C, when you press a key on your keyboard, the chord is replayed by the sequencer starting from the latter but keeping the same harmonic deviations (2 tones, 1.5 tones, 2.5 tones). Remember to activate tracking (TRACK), otherwise, the action on the keyboard will not be considered.

Parameter	Function
GATE	Advances to next step
RESET	Resets. Resumes at step 1
TRACK	Validates or disables pitch tracking at the pitch input
DIRECTION	Chooses the sequence direction
STEPS	Sets the number of steps
OFFSET	Defines the offset of the first step of the sequencer in relation to the sequence of notes in the sequence
PITCH line (C3, E3, G2, B2, etc.)	Allows you to choose the notes of each step, from C2 (C2) to C8 (C8)
GLIDE line	Enables or disables the glide for each step Gray: inactive Blue: pending Orange: in progress
GATE line	Gray square: note inactivated Blue square: note activated Orange square: currently activated note
GLIDE button	Glide length, from 0 to 2 s

Table 5.4. *Main parameters of the 8 STEPS module*

5.1.2.2. *Note and modulation sequencer*

We will incorporate a modulation sequencer into our rack in this second example. It is not the notes but the variations in modulations that modulate the filter or other elements (envelope, LFO, effects, etc.).

Here, we will modulate the parameters of the SVF filter block.

Instructions and comments are as follows:

– take the rack from the previous example; if you do not have it, create it;

– add the SEQ-BENTO 4 MODS SEQUENCER block from the Bento library;

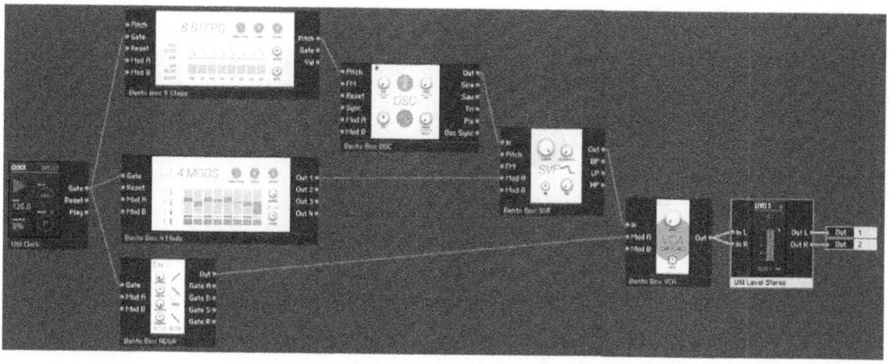

Figure 5.11. *Rack with a double sequencer*

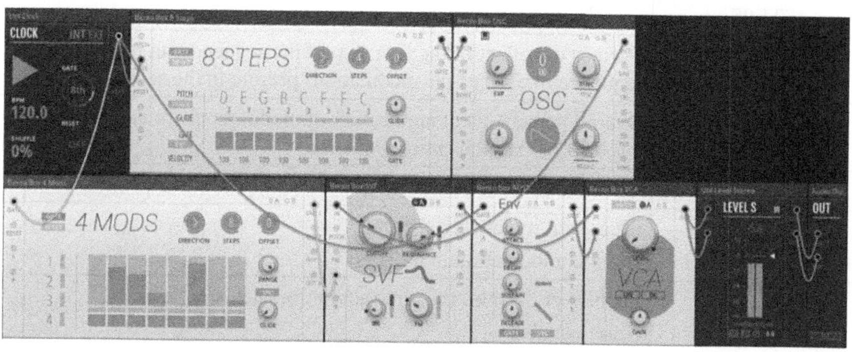

Figure 5.12. *Double sequencer rack in Panel mode*

– connect the GATE output of the CLOCK block to the GATE input of the 4 MODS block;

– connect the OUT 1 output of the 4 MODS block to the MOD A input of the SVF block;

– the valuable parameters of the 4 MODS block (Figure 5.13) for our exercise are described in Table 5.5. For more information, see the Native Instruments Reaktor Blocks documentation;

– try different modulation values and listen to how they work on the filter;

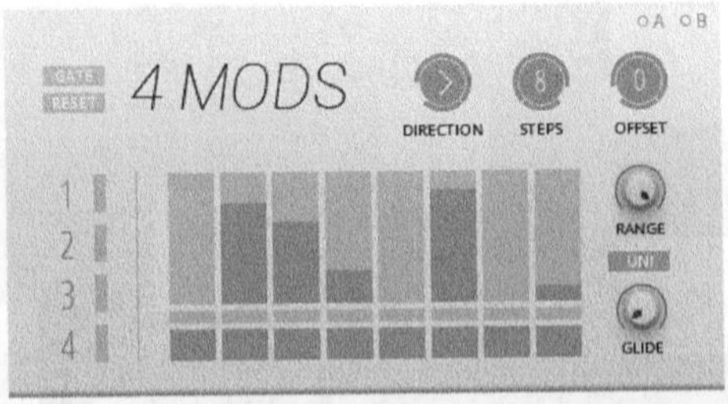

Figure 5.13. *The 4 MODS block from the Bento library*

Parameter	Function
DIRECTION	Chooses the direction
STEPS	Sets the number of steps
OFFSET	Defines the offset of the first step of the sequencer in relation to the sequence of each of the modulations present in the sequence
RANGE	Overall percentage of application of modulations for each step, from 0 to 100%, over the entire sequence
GLIDE button	Glide duration, 0–2 s
Modulation value line (upper line)	Blue: activity level (0–100%) Orange: modulation in progress
Glide line (middle line)	Enables or disables the glide action for each step Gray: inactive Blue: active Orange: in progress
Gate line (bottom line)	Gray rectangle: modulation inactivated Blue rectangle: modulation activated Orange rectangle: modulation in progress
1, 2, 3, 4	Selects output OUT 1, OUT 2, OUT 3 or OUT 4 relative to set modulation series 1, 2, 3 or 4

Table 5.5. *Parameters of the 4 MODS block*

– do not forget to configure channel A of the filter of the SVF block for CUTOFF, RESONANCE, and KEY TRACKING (see Figure 5.14) to activate the different modulations programmed in the 4 MODS block;

– modify the filter type by clicking on the curve to the right of the SVF acronym (high pass, band pass, low pass).

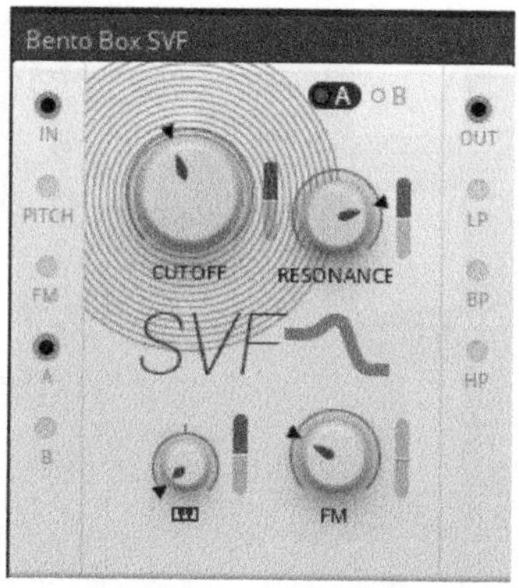

Figure 5.14. *Setting channel A of the SVF filter block*

5.1.2.3. *Arpeggiator*

In this exercise, we will create a rack with two VCOs, two envelope generators, a mixer, a filter and an arpeggiator, all controlled by a clock and amplified using a VCA.

The arpeggiator used will be the QUARP block. Here, we use version 1.1.

Instructions and comments are as follows:

– all VCO, VCF, GE and VCA blocks and mixers come from the Bento Box library (OSC × 2, SVF, ENV × 2, VCA and MIX);

– the QUARP arpeggiator block must be downloaded from the Reaktor user library on the Native Instruments website;

– connect the GATE output of the CLOCK block to the AUDIO/CLOCK IN input of the QUARP block;

– connect the RESET output of the CLOCK block to the RESET input of the QUARP block;

– connect the PITCH OUT output of the QUARP block to the PITCH inputs of the two OSC blocks and the SVF block;

– connect the GATE OUT output of the QUARP block to the GATE inputs of the two ENV blocks;

– connect the OUT output of each of the OSC blocks to the two inputs IN 1 and IN 2 of the MIX block;

– connect the OUT output of the MIX block to the IN input of the SVF block;

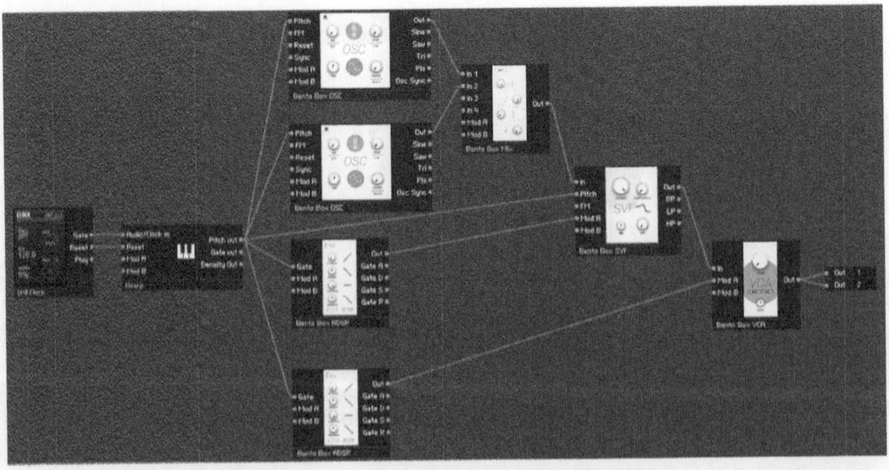

Figure 5.15. *Rack with arpeggiator*

Figure 5.16. *Arpeggiator rack in Panel mode*

– connect the OUT output of the first ENV block to the MOD A input of the SVF block;

– connect the OUT output of the second ENV block to the MOD A input of the VCA block;

– connect the OUT output of the SVF block to the IN input of the VCA block;

– connect the OUT output of the VCA block to the two OUT connection points of the OUT block;

– remember to configure channel A of the VCA block and configure volumes 1 and 2 on the MIX block;

– the parameters of the QUARP block are presented in Table 5.6;

Parameter	Function
ARPEGGIO	Defines the type of arpeggio (forward/backward, pendulum, random, etc.)
OCTAVE	Defines the number of octaves covered by the arpeggio
INVERT	Reverses the direction of the sequence
ARP	Arpeggio or quarp
Red square button	Stops the arpeggiator. A new chord must be played to resume
A–B	Selects modulation channel A or B

Table 5.6. *Settings of the QUARP block*

– set and start the clock and play a chord; the arpeggiator holds the arpeggio of the chord until the next chord is played;

– try the different arpeggiator settings, including different directions.

5.1.3. *Max/MSP*

For this exercise with Max/MSP, we will present two examples of sequencers and one example of an arpeggiator.

There are many options for creating this type of functionality; in the lines that follow, you will find some solutions. If the objects used are the same, the differences relate to the ergonomics of the controls made available to the user; the same goes for the arpeggiator.

5.1.3.1. *Simple sequencer*

This first sequencer offers 12 notes selectable by the user, using a linear pointer, between A0 to G9[1], that is, 10 octaves covered by the MIDI range.

1 See Appendix 4 for the complete list of MIDI notes with their numbers and frequencies.

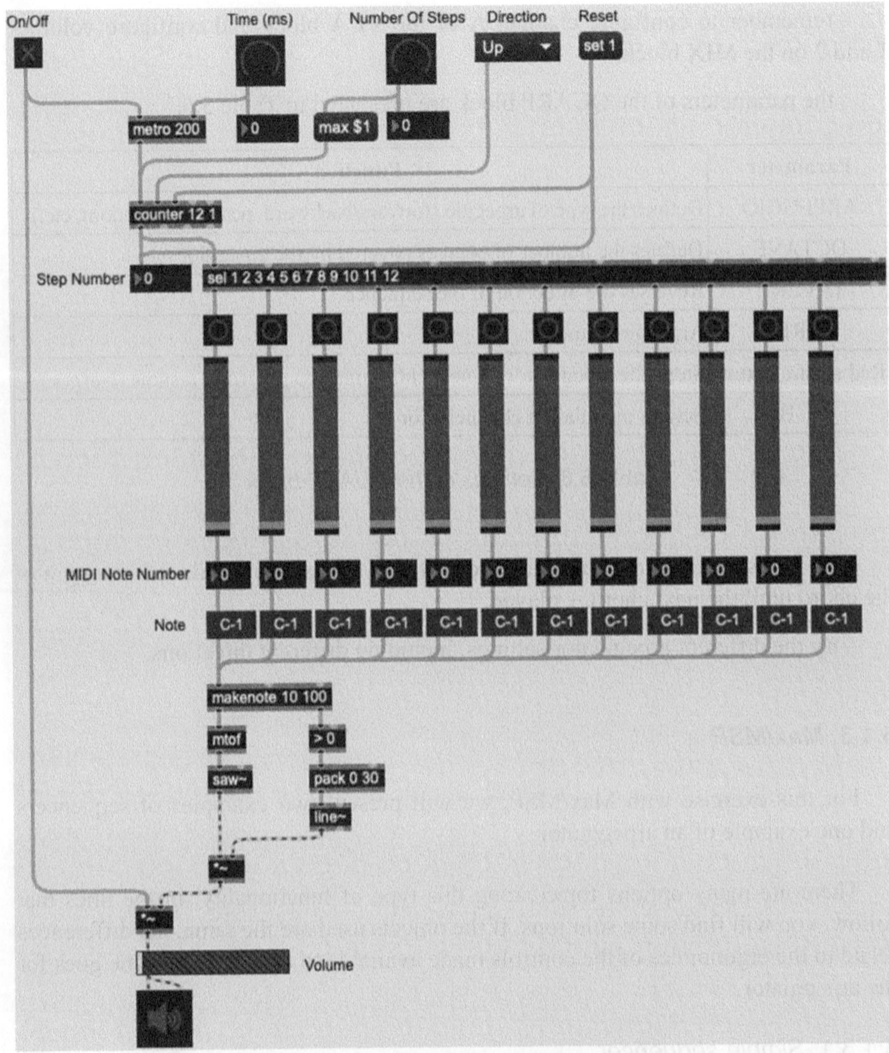

Figure 5.17. *Simple sequencer with Max/MSP*

The speed, or rather the time gap between each note, can be configured between 0 and 500 ms (very fast to very slow).

The number of steps can vary between 1 and 12, and three directions are proposed: upwards (Up), downwards (Down) and upwards then downwards (Up and Down).

A toggle allows you to activate the sequencer (On/Off), and a reset to step 1 is available (Reset).

A master volume control allows you to adjust the output level.

By default, a sawtooth signal is generated with a predefined envelope.

Instructions and comments are as follows:

– the output of a `toggle On/Off` object triggers a `metro 200` object and activates the passage of the audio signal via a `*~` object;

– the `dial Time (ms)` object defines the length of each step in milliseconds. Its size is equal to 401 and its minimum value is equal to 100. A `number` object placed at the output allows you to control the time value;

– the `dial Number Of Steps` object defines the number of steps. Its size is equal to 12 and its minimum value is equal to 1. A `number` object placed at the output allows you to control the number of steps;

– the `umenu Direction` object selects scroll direction (`Up`, `Down`, `Up & Down`);

– the `message set 1` object resets the step counter;

– the `metro 200` object receives the signal from the toggle On/Off object at its left input and at its right input the time from the output of the `dial Time (ms)` object. By default, the time is 200 ms;

– the output of the `metro 200` object is connected to the left input of the `counter 12 1` object;

– a `message max $1` object receives the number of steps as its input and returns the maximum number of steps as its output to the second input of the `counter 12 1` object;

– the third input of the `counter 12 1` object receives the `set 1` message, which resets the counter;

– the left output of the `counter 12 1` object is connected to the left input of a `sel 1 2...12` object, which determines the choice of its output according to the step received by the `counter 1 12` object;

– the `counter 12 1` object is initialized by default to a counting loop ranging from 1 to 12;

– a `number` object connected to the left output of the `counter 12 1` object controls the current step number;

– the 12 left outputs of the `sel 1 2...12` object are connected to a `bang` object, which indicates the activity of the current output;

– each of the outputs of the 12 `bang` objects is independently connected to the input of a `slider` object, which has a minimum output value of 21 and a size of 107;

– the output of each `slider` object is connected to the input of a `number` object to check the note number;

– the output of each `number` object is independently connected to the input of a umenu object, which displays the note from A0 to G9;

– each of the 12 `umenu` objects has the following occurrences: `C-1, C#-1, D-1, D# -1, E-1, F-1, F#-1, G-1, G#-1, A-1, A#-1, B-1, C0, C#0, D0, D#0, E0, F0, F#0, G0, G#0, A0, A#0, B0, C1, C#1, D1, D#1, E1, F1, F#1, G1, G#1, A1, A#1, B1, C2, C#2, D2, D#2, E2, F2, F#2, G2, G#2 A2, A#2, B2, C3, C#3, D3, D#3, E3, F3, F#3, G3, G#3, A3, A#3, B3, C4, C#4, D4, D#4, E4, F4, F#4, G4, G#4, A4, A#4, B4, C5, C#5, D5, D#5, E5, F5, F#5, G5, G#5, A5, A#5, B5, C6, C#6, D6, D#6, E6, F6, F#6, G6, G#6, A6, A#6, B6, C7, C#7, D7, D#7, E7, F7, F#7, G7, G#7, A7, A#7, B7, C8, C#8, D8, D#8, E8, F8, F#8, G8, G#8, A8, A#8, B8, C9, C#9, D9, D#9, E9, F9, F#9, G9;`

– the 12 outputs of the `umenu` objects are connected to the left input of a `makenote 10 100` object, which generates the defined MIDI note with a velocity of 10 and a duration of 100 ms;

– the left output of the `makenote 10 100` object is connected to a `mtof` object, which transforms the note into a frequency;

– the right output of the `makenote 10 100` object is connected to a `> 0` object, which validates the presence of a velocity to apply an envelope;

– the `saw~`, `*~`, `pack 0 30`, `line~`, `gain`, and `ezdac~` objects ensure signal generation and audio output. They have already been encountered in previous exercises;

– activate the sequencer, choose each note, the speed (time between each step), the number of steps and the sequence playback direction;

– remember to adjust the output volume to a level that suits you;

– a note can be chosen during the sequence.

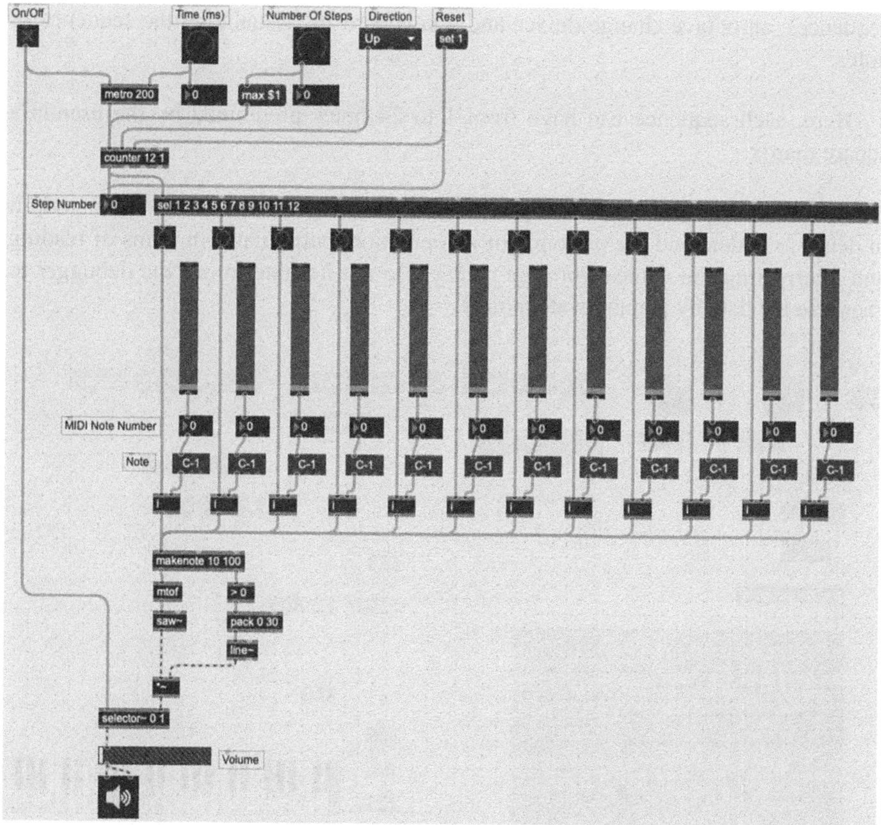

Figure 5.18. *Modifying the choice function of the sequencer notes*

NOTE.– When choosing a note during the sequence, we hear an unpleasant sliding of notes that disrupts the sequence. This comes from the fact that the data coming from the `sel 1 2...12` object via the `bang` object pass through the `slider` object to choose the notes. To overcome this problem, it is possible to divert the note choice function by adding an `i` object and modifying the connections as shown in Figure 5.18. The `i` object, which stores an integer value, retrieves the data transmitted by the `sel 1 2...12` object through the `bang` object and considers the choice of note.

5.1.3.2. *Advanced sequencer*

This second patcher presents a different sequencer to the previous one with more advanced features, such as recording presets and storing up six different note sequences, an octave change device and a keyboard for visualizing the sequence of notes.

Here, each sequence can have from 1 to 24 notes predefined by the user in a square matrix.

NOTE.– The programming of this patcher is quite complex; it is essential to study it in detail to understand the principle of its operation, particularly in terms of reading and interpreting the values from the matrix. Do not hesitate to use the debugger to visualize the data by placing watchpoints.

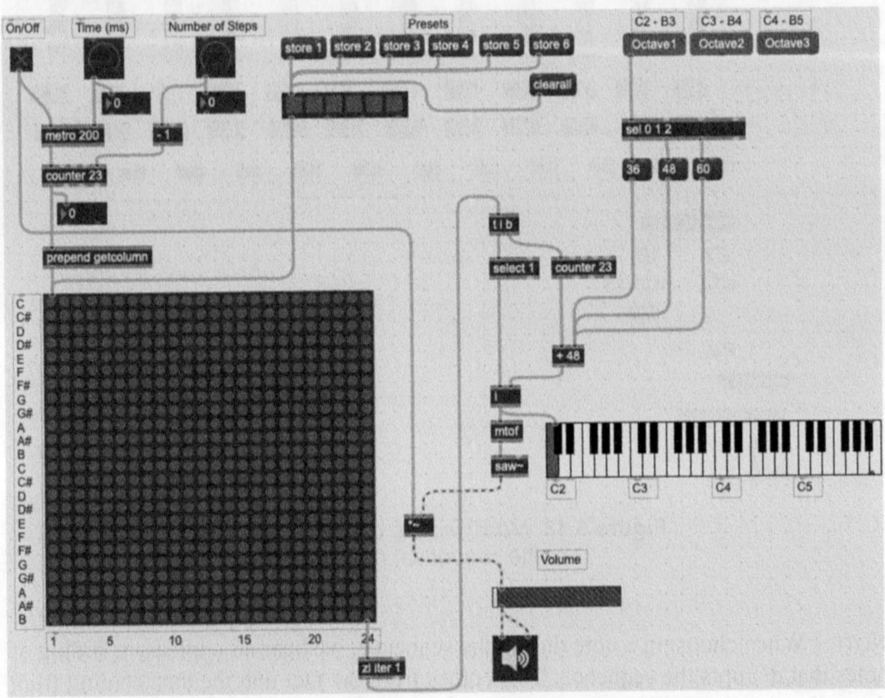

Figure 5.19. *Advanced sequencer with Max/MSP*

Instructions and comments are as follows:

– a `toggle` object triggers the activation of the sequencer. Its output activates the `metro 200` object and authorizes the output of the audio signal via the `*~` object placed in front of the `gain` object;

– a dial Time (ms) object determines the duration of each step. Its size is equal to 451 and its minimum value is 50. A number object, at the output, controls the chosen duration;

– the metro 200 object receives the value of the dial Time (ms) object at its right input. By default, the step duration is 200 ms;

– the output of the metro 200 object is connected to the counter 23 object, which defines the number of steps in the sequence, from 0 to 23, or a maximum of 24 steps. By default, the counter generates 24 steps;

– a dial Number Of Steps object defines the number of steps. Its maximum value is 24, and its minimum value is 1. A number object placed at the output allows you to control the number of steps;

– a - 1 object subtracts one from the value transmitted by the dial Number Of Steps object to transmit to the counter 23 object the correct number of steps (from 0 to 23) ;

– a number object connected to its left output displays the number of steps selected;

– the input of a prepend getcolumn object is connected to the left output of the counter 23 object to request the content of the column chosen by the counter 23 object by matrixctrl;

– the output of the prepend getcolumn object is connected to the input of the matrixctrl object;

– the matrixctrl object with 24 columns and 24 lines allows the user to define their choice of notes for the sequence;

– the output of the matrixctrl object is connected to the right input of a zl iter 1 object, which retrieves the values 0 (unchecked) or 1 (checked) one by one from the column previously requested by the prepend getcolumn object following the column number defined by the counter 23 object;

– the right output of the zl iter 1 object is connected to the input of a trigger t i b object, which triggers an integer value (i = 1) and the presence of a value (bang);

– the right output of the object t i b is connected to a select 1 object, which validates if the value received from the trigger t i b is 1;

– the left output of the t i b object is connected to the right input of a counter 23 object, which adds a looping counter on 23 values;

– an i object whose left input is connected to the left output of the select 1 object and the right input to the output of the + 48 object gets the value sent by the counter 23 object through the + 48 object. If the value 1 is detected by the select 1 object, a constant value of 48 is added to the value sent by the counter 23 object. This value of 48 determines the octave from which the sequence takes place (octave 2). At the output of the i object, we find the MIDI note number corresponding to the note in the matrix for the column currently being processed;

– the kslider object, which has a width of 48 notes (range) and an offset of 36 (lowest MIDI note), displays the number of the MIDI note at the output of the object i;

– the mtof object transforms the MIDI note number into a frequency; its output controls a sawtooth oscillator in the form of a saw~ object;

– the output of the saw~ object is directed towards a gain object followed by an ezdac~ object to control the volume and ensure audio output;

– 1 tab Octave 1 Octave 2 Octave 3 object provides three buttons that will allow the user to choose the octave;

– the output of the tab Octave 1 Octave 2 Octave 3 object is connected to the input of the sel 0 1 2 object whose outputs are activated depending on the button chosen (Octave 1 – output 0, Octave 2 – output 1, Octave 3 – output 2). Each output is connected to a message object;

– the three message 36, 48, and 60 objects have the MIDI note number of the octave, which is transmitted to the + 48 object. The value of 48 is the number of the MIDI note of the octave used by default;

– the output of each of the six message, store 1, store 2... store 6 objects each have their output connected to a preset object;

– the preset object has six preselects corresponding to the message objects in the previous point;

– a message clearall object connected to the input of the preset object allows you to erase all preselects;

– check the boxes in the matrix to generate the sequence that suits you. Avoid checking multiple boxes in the same column; in this case, only the lowest box in the matrix (highest note) will be considered. Click once on a box to check it, and again to uncheck it. To select boxes more quickly, you can drag your mouse over the matrix while pressing the left button. This is also valid for deleting them;

– set the step duration (time), number of steps and octave (1, 2 or 3);

– activate the sequencer (on/off) to start the sequence;

– do not forget to adjust the master volume;

– to record a sequence, click one of the store buttons;

– to recall a sequence, click one of the six boxes of the preset object;

– to clear all presets, click the clearall button;

– clicking on one of the store buttons already containing a preselection deletes it to replace it with the new one.

5.1.3.3. *Arpeggiator*

This arpeggiator uses functionalities not yet used in this work: the borax, coll, change and gate objects are new.

The sequence of arpeggio notes can be reversed via two Up and Down controls; the speed and overall volume are adjustable.

Its principle is based on the storage (coll) of a set of notes or a chord played by the user, which is then broadcast as an arpeggio, timed by a clock (metro).

Information related to note-on (key pressed) and note-off (key released) events generated by the user is captured by the borax object.

Instructions and comments are as follows:

– the mtof, tri~, *~, gain and ezdac~ are standard objects; we will not discuss their connections again;

– the notein object sends the MIDI note of the note and its velocity to the left input of a borax object;

– the borax object provides a set of information, including that of interest to us in this example, respectively, from left to right, a voice allocation number, the number of notes held, the pitch of the incoming note and the velocity of the incoming note (0 if no note enters);

– a pack 0 0 0 object retrieves the information from the previous point (allocation number, pitch, and velocity) from outputs 2, 4, and 5 (starting from the left) of the borax object to create a list. By default, the list contains 0 0 0;

– the left part of this patcher manages the recording of the arpeggio and the right part its playback;

– a > 0 object whose input is connected to output 5 of the borax object tests the velocity value, i.e., whether a note is pressed;

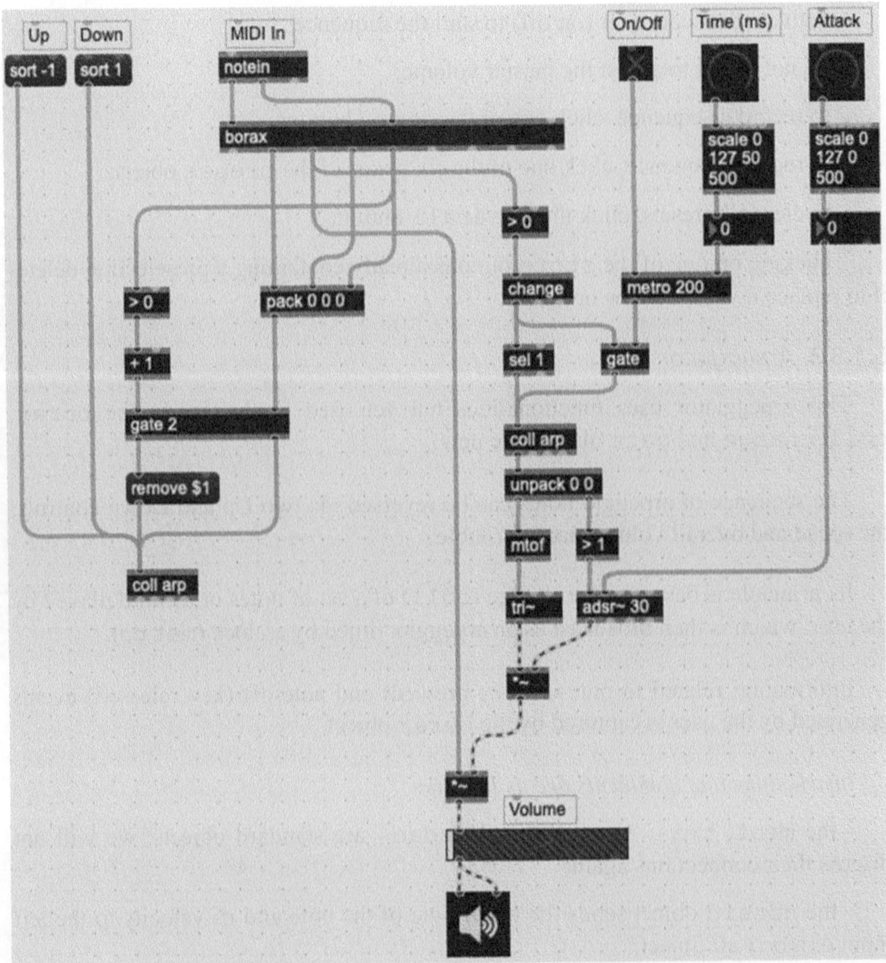

Figure 5.20. *Arpeggiator with Max/MSP*

– the output of the `> 0` object is connected to the input of a `+ 1` object, which generates a value of 2 (1 + 1) in the event of a positive test (velocity > 0, key pressed);

– the output of the `+ 1` object is connected to a `gate 2` object, which directs the incoming data (2 or the list) created by the object `pack` toward its left and right output;

– the output of the `pack 0 0 0` object is connected to the right input of the `gate 2` object;

– if the value received by the `gate 2` object at its left input is equal to 2, then it is transmitted to the input of a `remove $1` message object, which removes the note of the key that has just been released from the arpeggio;

– if the value received by the `gate 2` object is not 2, then the list created by the `pack 0 0 0` object is transmitted to the input of the `coll arp` object;

– the `coll arp` object receives three pieces of information at its input: a descending sort order delivered by the output of a `message sort - 1` object, an ascending sort order delivered by the output of a `message sort 1` object, or the list created by the `pack 0 0 0` object. This information is recorded in the `arp` data collection;

– the `coll arp` object saves the sequence of notes (chord played) and their different velocities from the list or removes (`remove $1`) from the sequence the note with a velocity of 0 (key released);

– Figure 5.21 shows an example of content saved in the `arp` data collection. We can see the four notes of a Cmaj (C major) chord with their voice numbers, followed by the MIDI number of each of the four notes and their velocity;

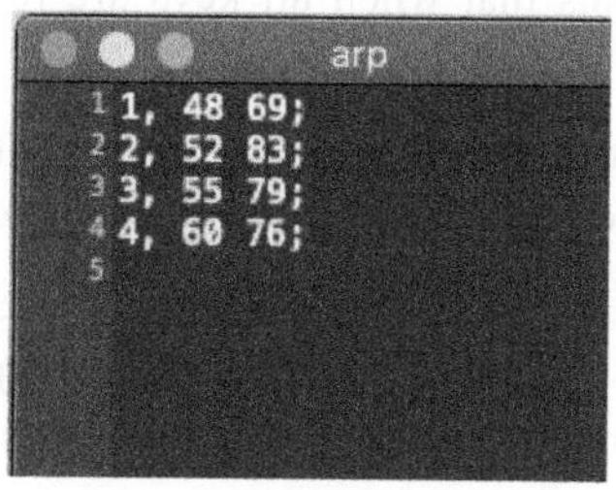

Figure 5.21. *The `arp` data collection (Cmaj chord)*

– now let us move on to the reading part, located on the right;

– a `toggle` object activates or stops the arpeggiator;

– a `dial` object, followed by a `scale 0 127 50 500` object, generates the time value of the arpeggio step, between 50 and 500 ms;

– the `number` object, placed at the output of the `scale 0 127 50 500` object, verifies the duration chosen by the user;

– a `metro 200` object whose left input is connected to the output of the `toggle` object and the right input to the output of the `number` object sets the cadence of the arpeggio. By default, it is 200;

– a > 0 object connected to the third output from the left of the borax object retrieves the number of keys pressed and verifies if it is greater than 0; in this case, it transmits the value 1 to the change object;

– the output of the object > 0 is connected to the input of a change object, which checks if there has been a change (change from 0 to 1 – note pressed);

– the left output of the change object is connected to the inputs of a sel 1 and a gate object;

– the sel 1 object sends a bang object if it receives a value of 1 at its left input from the change object;

– the left input of the gate object is connected to the left output of the change object;

– the right input of the gate object is connected to the output of the metro 200 object;

– if the gate object receives a value of 1 from the change object, it lets the clock signal from the object pass to its output metro 200; otherwise, the clock signal is blocked. This means that when all keys are released, the clock signal is blocked;

– the coll arp object transmits all arpeggio notes to its output, including for each, their voice number, the MIDI number of the note and its velocity;

– the left output of the coll arp object is connected to an unpack 0 0 object, which breaks down the data from the coll arp object into a MIDI note number sent to its left output and a velocity sent to its right output;

– the > 1 object receives the velocity of each note from the right output of the unpack 0 0 object. If it is greater than 1, then it is transmitted to the adsr~ 30 object, which defines an envelope whose attack is 30 ms by default;

– the two rotating dials dial Time (ms) and Attack allow the user to choose the arpeggio step duration and the note attack duration (envelope);

– the two dial objects have a default value range (0 to 127). A scale 0 127 50 500 object placed at the output of each of them transforms the range of values (0 to 127) into a range of values from 0 to 500, in this case milliseconds;

– both number objects placed behind the scale 0 127 50 500 objects control user-selected values for duration and attack;

– the output of the number object placed to the left is connected to the right input of the metro 200 object;

– the output of the `number` object placed to the right is connected to the second left input of the `adsr~ 30` object to configure the attack. By default, the latter is 30 ms;

– turn on the arpeggiator (on/off), set a step duration and an attack duration. Make sure that the attack duration is less than the step duration, otherwise, it will not be fully considered as the step duration is shorter than that of the attack;

– play a chord on the MIDI keyboard; you should hear the arpeggio;

– by clicking the Up and Down buttons, the arpeggio notes can be sorted in ascending or descending order;

– remember to set an appropriate output volume;

– when you release the keyboard keys, the arpeggio stops.

NOTE.– As with the previous patcher, programming the arpeggiator is complex. Study the function of its different objects to understand how it works, particularly in managing the `arp` collection (`coll arp`) in writing and reading. Do not hesitate to place watchpoints and use the debugger.

5.1.4. *Pure Data*

In this section, we will successively create a simple sequencer very close to the one we have already created with Max/MSP (see section 5.1.3.1) and an arpeggiator with a graphic display capable of capturing chords of 2 to 5 notes.

5.1.4.1. *Sequencer*

This Pure Data sequencer works on 12 steps with MIDI notes, varying between 21 (A0 – Do0) and 127 (G9 – Sol9)[2].

The user adjusts the speed of the sequence, the number of steps, the attack of the notes and the general volume.

Instructions and comments are as follows:

– a `toggle` object activates or deactivates the sequencer and audio output;

– a `slider Time (ms)` object covering a range of linear values between 50 and 1,000 manages the step duration;

---

2 See Appendix 4 for the complete list of MIDI notes with their numbers and frequencies.

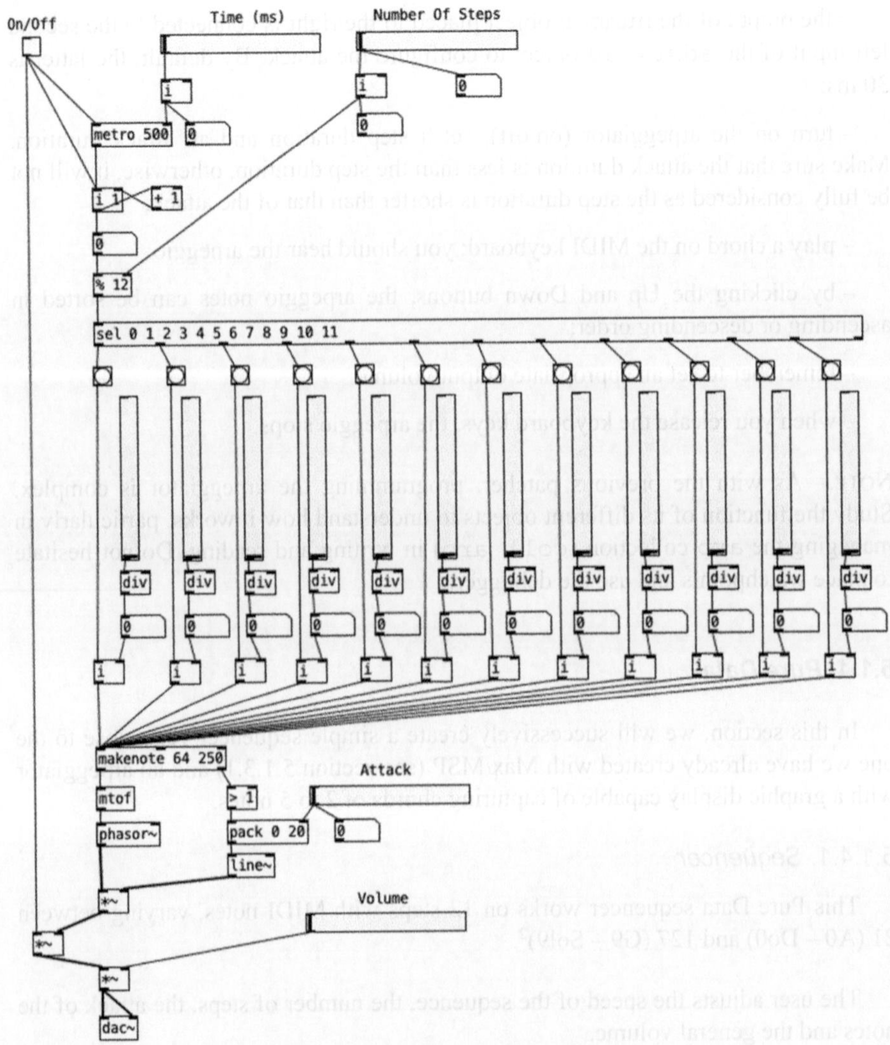

Figure 5.22. *Sequencer with Pure Data*

– a `slider Number Of Steps` object covering a range of linear values between 1 and 12 manages the number of steps;

– both `i` objects connected to the output of the `slider` object remove any decimal values to keep only the integer;

– the output of the left `i` object is connected to the right input of a `metro 500` object as well as the input of a `number` object, which displays the chosen duration;

– the output of the right `i` object is connected to the right input of a `% 12` object as well as the input of a `number` object, which displays the step number;

– the left input of the `metro 500` object is connected to the `toggle On/Off` object;

– the `metro 500` object serves as a clock to regulate the sequencer. By default, the duration of a step is 500 ms;

– the `i 1` and `+ 1` objects form a counter to increase the sequencer steps by 1;

– the left input of the object `i 1` is connected to the output of the `toggle On/Off` object and to the left output of the `metro 500` object;

– the right input of the object `i 1` is connected to the output of the object `+ 1`;

– the output of the object `i 1` is connected to the left output of the object `+ 1` and to a `number` object, which displays the number of steps since the start of the sequence. This number is reset to 0 each time the sequencer is launched;

– the output of the `number` object is connected to the input of the object `% 12` and calculates the current step number between 0 and 11;

– the output of the object `% 12` is connected to the input of a `sel 0 1...11` object, which selects, one after the other, one of the steps in the sequence;

– the right input of the object `% 12` receives the number of steps the user selects. By default, it is 12;

– all outputs of the `sel 0 1...11` object are connected to the input of a `bang` object, which indicates the current step;

– all outputs of the `bang` objects are connected to the left input of an `i` object, which contains the MIDI note number of the current step;

– for each of the 12 steps, a `slider` object, a `div` object, and a `number` object allow you to choose the number of the MIDI note. The `div` object only keeps the integer part of the number emitted by the `slider` object. The `number` object displays the MIDI note number, from 21 to 127[3];

– the output of each of the 12 `number` objects is connected to the right input of an `i` object;

---

3 See Appendix 4 for the complete list of MIDI notes with their numbers and frequencies.

– the output of each of the 12 `i` objects is connected to the left input of a `makenote 64 250` object, which generates the defined MIDI note with a velocity of 64 and a duration of 250 ms by default;

– the left output of the `makenote 64 250` object is linked to a `mtof` object, which transforms the MIDI note number into a frequency;

– the output of the `mtof` object is connected to the input of a `phasor~` object, which generates a sawtooth signal;

– the right output of the `makenote 64 250` object is linked to a `> 1` object which verifies if the velocity is greater than 1; in this case, it passes it to a `pack 0 20` object, which creates a list for the duration of the attack;

– the right input of the `pack 0 20` object retrieves the duration of the attack at the output of the `slider Attack` object, which covers a range of linear values from 10 to 500 ms;

– a `number` object connected to the output of the `slider Attack` object displays the attack duration;

– the output of the `pack 0 20` object is connected to the input of a `line~` object, which generates the attack for the signal envelope (ascending ramp from 0 to 500 maximum);

– the three `*~`, the `slider volume` and the audio output `dac~` objects have already been encountered in other exercises. They control the volume and output the audio signal, which is not present when the sequencer is inactive;

– optionally adjust the duration of each step (sequencer speed), the number of steps, the attack time and the overall volume;

– the attack time must remain less than or equal to the step duration, otherwise it will be cut and signal attenuation will occur.

5.1.4.2. *Arpeggiator*

This arpeggiator is designed around a table in which the different notes of the chord are stored. The number of notes can be selected – between one and five – and the speed (duration) of the sequence of notes can be configured according to a varying interval between 50 and 500 ms.

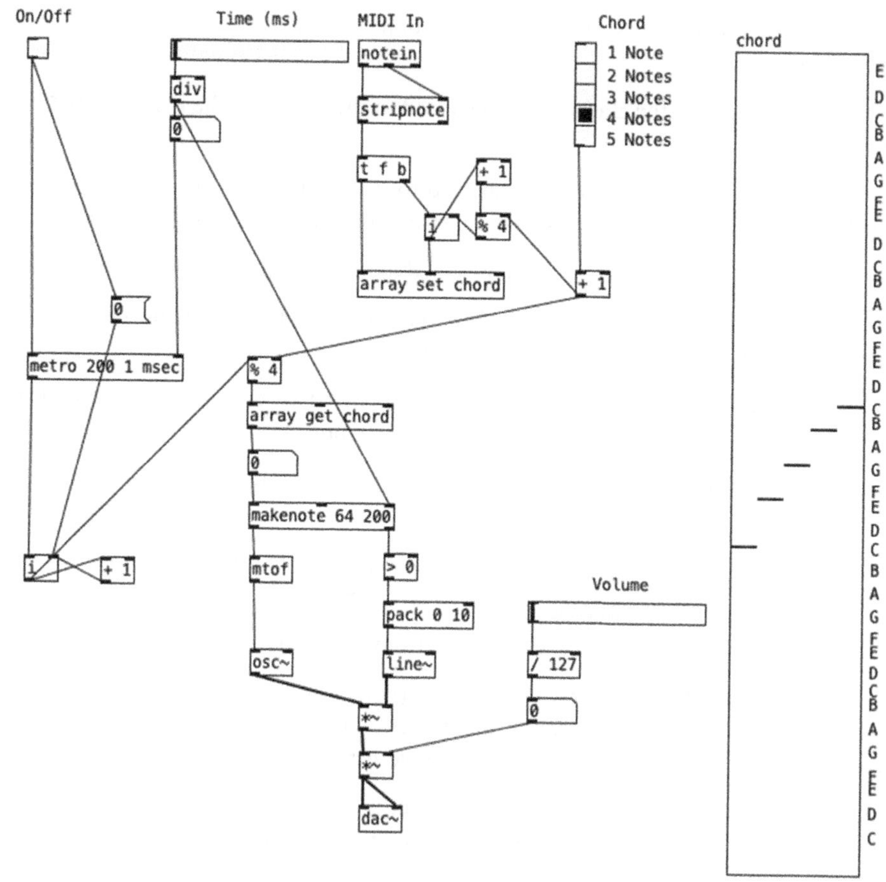

Figure 5.23. *Arpeggiator with Pure Data*

Instructions and comments are as follows:

– a `toggle On/Off` object activates or disables the arpeggiator. It initializes the note counter and starts the clock that ticks the arpeggio;

– the output of the `toggle On/Off` object is connected to the input of a `metro 200 1 msec` object, which manages the time interval between the different arpeggio notes and at the arrival of a `message 0` object, which resets the counter each time the arpeggiator is activated;

– the output of the `message 0` object is connected to the right input of the `f` object;

– the note counter is designed with an `f` object and a `+ 1` object that increases the value;

– the left input of the `f` object is connected to the output of the `metro 200 1 msec` object;

– the output of the `f` object is connected to the left input of the `+ 1` object and the left input of a `% 4` object, which defines an arpeggio of four notes by default;

– the output of the object `+ 1` is connected to the right input of the object `f`;

– a `slider Time (ms)` object sets the duration of the arpeggio intervals. It covers a range of linear values from 50 to 500;

– the `slider Time (ms)` output is connected to the left input of a `div` object, which will keep the value from the `slider` object;

– the output of the `div` object is linked to a `number` object, which will display the chosen duration and to the right input of the `makenote 64 500` object;

– the output of the `number` object is connected to the right input of the `metro 200 1 msec` object;

– a `notein` object captures MIDI notes from the keyboard;

– the left (MIDI note number) and central (velocity) outputs are connected to the left and right inputs of a `stripnote` object, which prevents a note from being considered when releasing a keyboard key (note-off);

– the left output of the `stripnote` object is connected to the left input of a `t f b` object, which triggers the transmission of a note at its right output and validates the presence of a velocity at its left output;

– objects `i`, `% 4`, and `+ 1` manage the arpeggio note counter;

– the right output of the `t f b` object is connected to the left input of object `i`;

– the output of object `i` is connected to the central input of an `array set chord` object and at the left input of a `+ 1` object;

– the output of the `+ 1` object is connected to the left input of the `% 4` object;

– the output of the `% 4` object is connected to the right input of the `i` object;

– the `array set chord` object saves note values to a table called `chord`. Its left input retrieves the MIDI note number and its right input indexes the latter in the table. Figure 5.24 shows an example of the chord array contents, here a Do7 (C7) chord;

– a `vradio Chord` object with cells (0 to 4) allows you to select the number of notes in the arpeggio between 1 and 5;

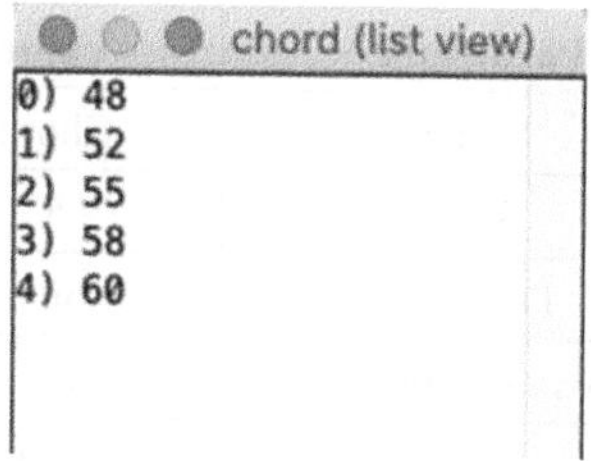

Figure 5.24. *Example of content from the* `chord` *table*

– its output is connected to a `+ 1` object, which increases the output values of the `vradio Chord` object so that they are between 1 and 5;

– do not forget to initialize the `vradio` object on notes position 4 (Parameters Init) so that the patch opens by default with 4 notes checked;

– the output of the `+ 1` object is connected to the two right inputs of the `% 4` objects to transmit the number of notes chosen by the user;

– the output of the lower `% 4` object is connected to the left input of an `array get chord` object to read the list of recorded notes;

– a `number` object connected to the output of the `array get chord` object allows you to check the number of each read note;

– the output of the `number` object is connected to the left input of a `makenote 64 200` object, which retrieves a duration identical to that transmitted to the clock (`metro 200 1 msec`) to synchronize the cadence of the arpeggio and ensure each note is held;

– the right input of the `makenote 64 200` object is connected to the output of the `div` object placed behind the `slider Time(ms)` object;

– the left input of the `makenote 64 200` object is connected to the output of the `number` object to retrieve the read MIDI note. The velocity will always be 64, and the default duration is set to 500 ms;

– the `chord` table has the characteristics of Table 5.7; the parameters that are not mentioned keep their default value;

– if the `Save contents` option is checked in the `chord` table properties, the last arpeggio is retained even after the patch is closed;

– all the objects described below have a configuration that has already been encountered in previous chapters, for example in section 2.2.4;

Parameter	Value
Name	Chord
Size	5
Draw as	Points
X range: from	0–5
Size	100
Y range: from	90–20
Size	600

Table 5.7. *Characteristics of the* `chord` *table*

– the `mtof` and `osc~` objects generate a sinusoidal signal corresponding to the notes of the arpeggio;

– the objects `> 0`, `pack 0 10` and `line ~` and `*~` create and combine the envelope of the audio signal;

– the `*~`, `slider Volume`, `/ 127`, `*~` and `dac~` objects provide audio output and volume management;

– turn on the arpeggiator (on/off), adjust the step duration (`Time ms`), the number of notes in the arpeggio and the overall volume.

NOTE.– Rather than playing the chord's notes to be arpeggiated on the keyboard, it is also possible to directly enter the notes into the table using the mouse, but this is not very precise.

5.2. Conclusion

In this chapter, we have looked at two tools that are often available on both vintage and contemporary synthesizers. Since the beginning of musical electronics, the sequencer and the arpeggiator have been systems that have always been available. Their usefulness may seem to fall within the realm of toys or gadgets. Still, many people have used them in very different musical styles from psychedelic, repetitive, contemporary, new wave, krautrock, electro-acoustic, etc.

Raymond Scott, Pink Floyd, Tangerine Dream, Klaus Schulze, Kraftwerk, Depeche Mode, Steve Reich, Philip Glass, and Pierre Boulez are examples of bands or musicians who have used these technologies.

Software like Ableton Live, Cubase, Logic Pro, FL Studio, Pro Tools, Reaper, Reason, Sonar, Digital Performer and many others are nothing more and nothing less than evolved versions of the first sequencers and arpeggiators; their functionalities are very advanced, and they allow you to do many other things, but the basic principles remain the same.

Conclusion

After two volumes and several hundred pages, this book is complete. If you have worked through both books, I hope you have acquired the fundamentals or reinforced your knowledge of subtractive sound synthesis and its associated synthesizers.

Although this work is voluminous, it is only the foundations. The field of sound synthesis is vast, and it is in constant evolution; the imagination, inventiveness and genius of inventors, engineers and researchers know no limits. As for artists, musicians, composers or sound designers, their thirst for new functionalities makes them the drivers of innovation, constantly pushing the boundaries of the impossible in search of the ultimate sound rendering in a continually expanding musical universe.

Even though subtractive synthesis is one of the oldest techniques regarding synthesizers, it is far from being outdated, and many machines still use it today. Its sound reproduction is incomparable and a must if you are looking for a loud sound.

Since the beginning of the 1990s, with the rise of digital technologies, synthesizers based on subtractive synthesis have become much more powerful, investing in MIDI without neglecting the founding CV/Gate standard. They have become communicating machines that can be easily interfaced with other systems, controlled by microcomputers, sequencers and groove machines, and interact with many other peripherals.

Not only has sound synthesis hardware continued to evolve but sound synthesis software has become omnipresent in recent decades. Several hundred virtual synthesizers are now available in the form of plugins, standalone versions or specific applications for microcomputers, tablets and even smartphones. They span the entire field of musical computing. You can find everything and anything, and it is very difficult to trace your path among so many elaborate products with functions so advanced that they can seem absurd. Despite this, a few flagship programs are unanimous, and their reputation is well established; you saw a few of them in this book.

Virtual modular synthesizers like the VCV Rack or Native Instruments Reaktor Blocks are excellent workarounds to their hardware counterparts. They make it easy to create high-performing assemblies but may cause frustration when turning knobs or switches from a screen. Nothing replaces the feeling of sliding or twisting a control with your fingertips, but let us face it, the software costs are much less than the hardware equivalent, and testing sophisticated configurations this way is much easier and more affordable.

For the most ambitious and passionate, visual sound programming languages like Pure Data or Max/MSP have acquired a maturity that opens unlimited perspectives in terms of invention. Everything becomes possible; only the imagination and skill of the developer define the limits of what is achievable. Personally, they are great educational tools for discovering sound synthesis. They provide detailed knowledge of the mechanisms and basics necessary for building and understanding how a synthesizer works.

What will the trends be in the coming years?

The answer to this question is complicated, but it is easier to understand after walking the aisles of the Superbooth in Berlin or the NAMM in Anaheim, which took place just as I wrote these lines.

As always, there have been many new developments in traditional violin making and digital electronics; however, we do not see anything truly revolutionary, and what is striking is if we look at the past trend.

For some time now, many manufacturers have started building analog synthesizers using standard techniques or even often improved replicas of analog synthesizers from the 1970s to 1980s. Here, innovation lacks imagination!

Most integrate MIDI or USB ports and, very often, digital command control, allowing for more flexible management and editing of parameters using a microcomputer.

In short, analog is in fashion; its grain, roundness, warmth and depth of sound appeal to many professionals. Its very organic sound reproduction contrasts with digital systems, which are colder and more crystalline. Even studio equipment is paying the price, reissuing old compressors, equalizers, preamps, etc.

If there are no limits from a financial aspect, why not invest in a vintage machine? They have never been so trendy, and it is still possible to find a rare gem. Their Spartan side and their authenticity make them instruments of choice. Although many of them are very robust, you should be wary of specific models, which, even when well maintained, are not immune to breakdowns that are often difficult to repair, as some components are no longer available and difficult to replace.

With the appearance and generalization of the MIDI 2.0 standard and its new functions, we will perhaps witness developments that could reshuffle the cards, bringing many new features within the small world of musical electronics. Since January 2020, this new protocol has been available, but its implementation and use have not yet become widespread. It should not be too long before its popularization intensifies.

Over time, synthesizers have become part of all forms of sound synthesis, including those based on:

– abstract algorithms such as subtractive synthesis, frequency modulation (FM) synthesis, waveshaping, phase distortion and pulse width modulation (PWM);

– sound samples such as granular synthesis or wavetable synthesis;

– signal models such as additive synthesis and analog modeling synthesis;

– physical modeling such as digital waveguide synthesis and modal synthesis.

Despite this wide array of techniques, most synthesizer manufacturers have retained subtractive synthesis, FM synthesis, analog modeling synthesis and wavetable synthesis, with others being present anecdotally.

Today, the number of synthesizers, hardware or software present on the market is vast. We will surely witness, as we saw in the 1990s, a refinement of reference models.

In any case, the course of time and its technological developments have not lessened the use of subtractive sound synthesis within synthesizers and we remain convinced that it will still be present in the decades to come to the delight of electronic music enthusiasts.

Appendix 1

USB Connectivity

USB (Universal Serial Bus) is a connection port intended for computer peripherals, with version 1.0 dating back to 1996.

Over time, we saw the appearance of versions 1.1, 2.0, 3.0, 3.1 and 4, with the main improvement being the throughput:

– USB 1.0: the theoretical version that defined the specification;

– USB 1.1 (1998): 1.5 or 12 Mbps;

– USB 2.0 (2000): 480 Mbps;

– USB 3.0 (2008): 5 Gbps;

– USB 3.1 (2013): 10 Gbps;

– USB 3.2 (2015): 20 Gbps;

– USB4 (2017): 40 Gbps and Thunderbolt 3 features (40 Gbps).

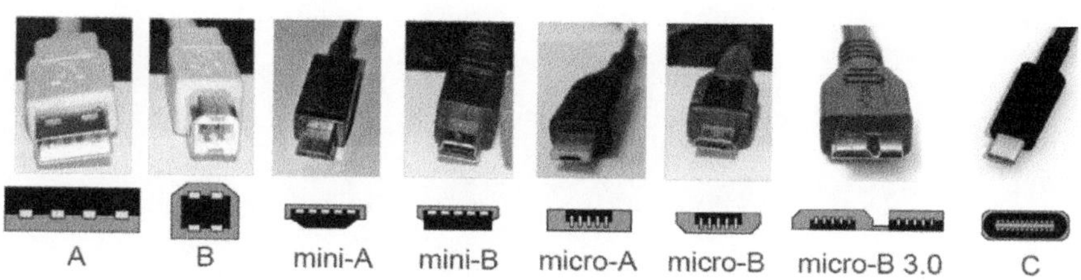

Figure A1.1. *The most common types of USB connectors. For a color version of this figure, see www.iste.co.uk/reveillac/synthesizers2.zip*

NOTE.– Over time, the standards have been renamed by the USB Implementation Forum (USB-IF):

– USB 3.0 is also called USB 3.1 Gen 1 or USB 3.2 Gen 1;

– USB 3.1 is called USB 3.1 Gen 2 or USB 3.2 Gen 2;

– USB 3.2 has become USB 3.2 Gen 2.

Connections have also evolved with different connectors: type A, type B, mini A, mini B, mini AB, micro A, micro B, micro AB and type C.

Appendix 2

Pure Data Extensions

Pure Data offers many features and the fact that it is open source, and therefore free, means the software is used by many developers. Indeed, it is less accomplished than its Max/MSP counterpart, and its interface is not as friendly, but in terms of programming, it is undoubtedly a rival. Did they not both have Miller Puckette as their original author?

On the one hand, it is more intuitive and more open. On the other hand, it benefits from a hard-working and responsive community of developers (see the reference section at the end of this book).

In this appendix, you will find tools and tips to help you work more efficiently with Pure Data.

A2.1. Oscilloscope

It can be helpful to have an oscilloscope to visualize a signal within a Pure Data patch. The abstraction described below allows you to build a simple but very useful version.

This oscilloscope has an input, a sensitivity setting and a time base setting.

Instructions and comments are as follows:

– create a new patch and implement the objects present in Figure A2.1;

– create an `array` object. Its properties are shown in Figure A2.2;

For a color version of all the figures in this chapter, see www.iste.co.uk/reveillac/synthesizers2.zip.

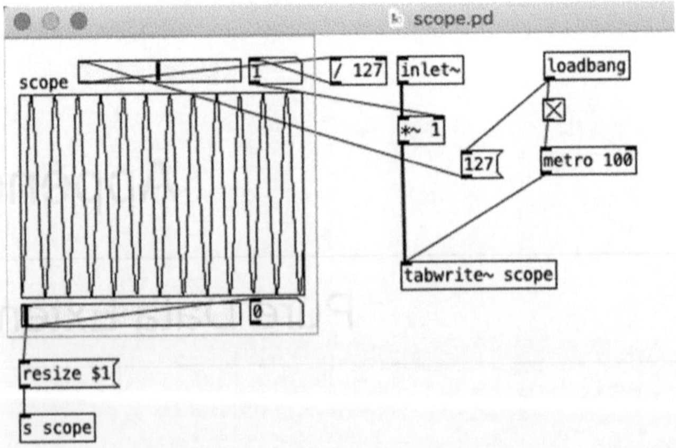

Figure A2.1. *Oscilloscope in Pure Data*

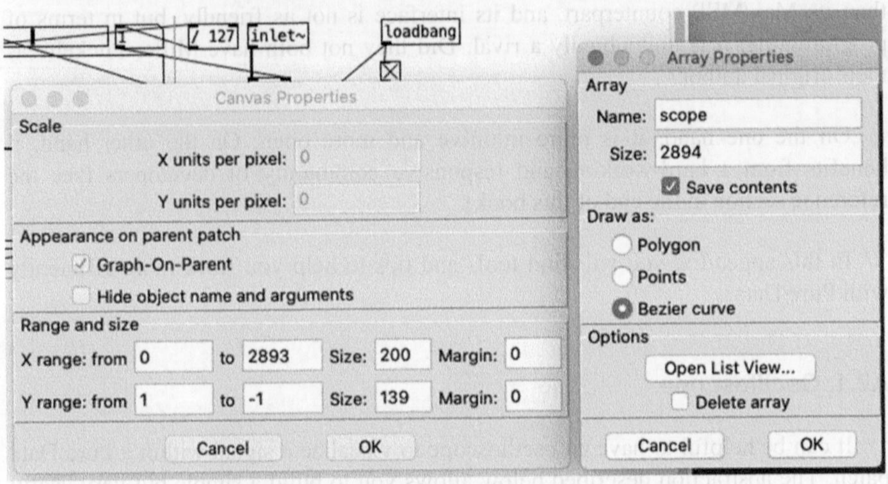

Figure A2.2. *Properties of the* `array` *object*

– an `inlet~` object allows the signal input. Its output is connected to a `*~ 1` object, which combines the signal received with the sensitivity level as the input;

– connect the output of the `*~ 1` object to the input of the `tabwrite~ scope` object;

– a `loadbang` object initializes the clock and the sensitivity value of the oscilloscope. Its output is connected to a `toggle` object and a `message 127` object;

– a `toggle` object starts the clock. Its output is connected to the left input of a `metro 100` object, which acts as a clock. By default, it clocks at 100 ms;

– the output of the `metro 100` object is linked to the input of the `tabwrite~ scope` object, which transcribes the received data into a `scope` array;

– the input of the `slider` object (above the signal display area) is connected to the output of the `message 127` object and the output of the `slider` object to the left input of the `/ 127` object;

– the `slider` object covers a linear value range from 0 to 254 and its width is 110 pixels. The other settings are left at their default values;

– the output of the `/ 127` object is connected to the input of the `number` object, which displays the sensitivity from 0 to 2;

– the `/ 127` object divides the `slider` sensitivity value so it is between 0 and 2;

– a `slider` object (below the signal display area) covers a range from 20 to 5000. It acts as the oscilloscope time base setting. Its width is 160 pixels. The other settings are left at their default values;

– the input of the `number` object is connected to the output of the time base `slider` object and controls the slider value chosen;

– the input of the `message resize $1` object is connected to the output of the time base `slider` object;

– the `resize $1` object resizes the signal display and, therefore, the `scope` table content;

– the output of the `resize $1` object is connected to a `scope` object, which transmits the values to the `scope` table;

– set the `canvas` properties of the patch as shown in Figure A2.3;

– save the patch as "scope.pd". It will become the `scope` abstraction;

– create a new patch;

– insert an `osc~ 440` object;

– insert a `scope` object;

– connect the output of the `osc~ 440` object to the left input of the `scope` object. The curve of a sinusoidal signal should appear in the frame of the `scope` object;

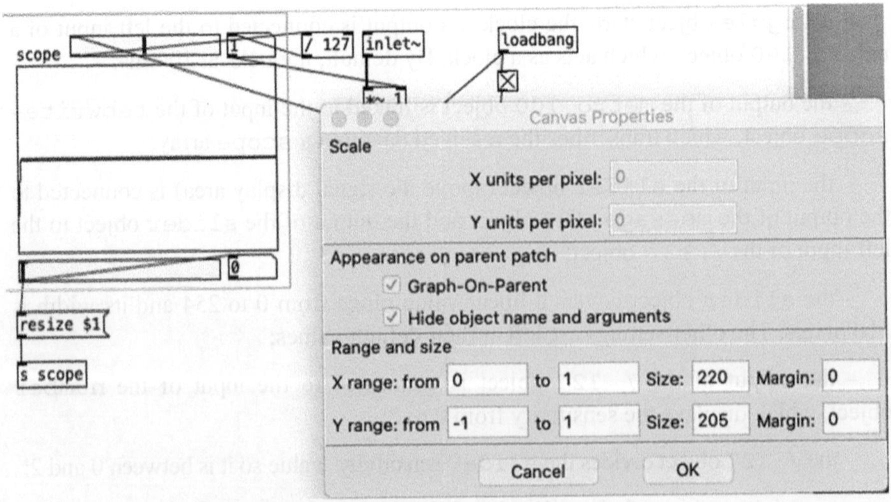

Figure A2.3. *Canvas proprieties of the patch*

– if the scope object does not appear, save this new patch (for example, as "test.pd") in the same directory as your scope patch. Close Pure Data and reload your patch;

– the slider at the top adjusts the sensitivity (amplitude of the signal curve), and the slider at the bottom adjusts the time base (frequency of the signal curve).

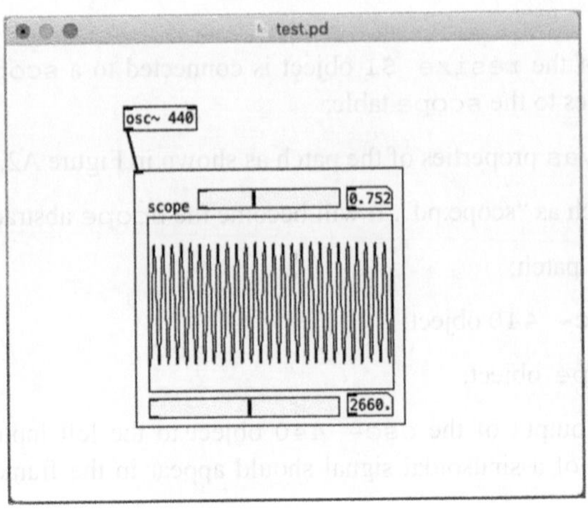

Figure A2.4. *Patch test with the scope object*

If you are looking for a more sophisticated oscilloscope, the `pad.scope.pd` abstraction developed by Pierre Guillot is downloadable for free and is an excellent alternative. You will find the download link in the reference section of this book.

A2.2. Activating/deactivating the DSP

Activating the DSP directly from a patch without clicking in the Pure Data Pd window or the Media menu can be interesting and practical.

When the DSP is inactive, no audio signal is available at the output, and Pure Data remains muted (no signal processing).

You can activate it using two message objects in your patch.

Figure A2.5. *DSP activation/deactivation message*

A2.3. Virtual keyboard

When a MIDI keyboard is unavailable, using your computer keyboard or a virtual one as a musical keyboard may be beneficial.

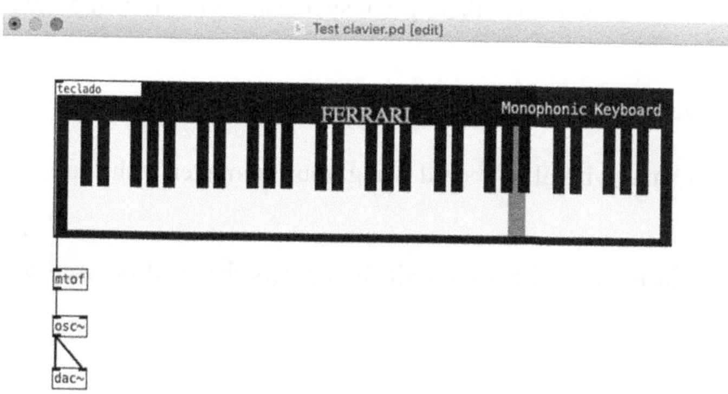

Figure A2.6. *An example using a virtual keyboard (teclado.pd)*

There are several solutions:

– download a free Pure Data abstraction, available on the Internet, such as the teclado.pd, keybin.pd and fr4-azerty.pd patches. Internet download links are given in the reference section of this book, in the "Virtual keyboard" section;

– install a virtual keyboard on your computer for your operating system. The Virtual MIDI Piano Keyboard (VMPK) software available for Microsoft Windows, Apple macOS, and Linux is a great solution. This open-source software, developed by Pedro López-Cabanillas, is available free of charge, and its download link is also available in the reference section of this book;

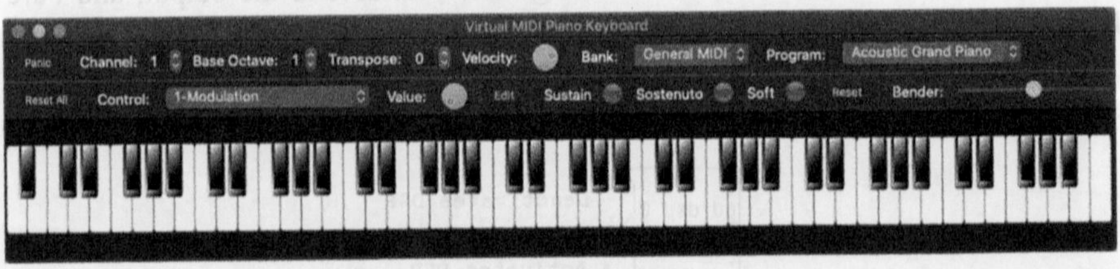

Figure A2.7. *VMPK, a virtual MIDI keyboard for a microcomputer*

– create your own virtual keyboard abstraction with Pure Data, as explained in section A2.3.1.

NOTE.– The Virtual MIDI Piano Keyboard is not exclusively reserved for Pure Data. It can be used as a software keyboard with Max/MSP, the VCV Rack or Native Instruments Reaktor and many other virtual synthesizers or musical programs.

A2.4. A virtual keyboard patch

Creating a small virtual keyboard using your computer with Pure Data is quite simple.

The patch in Figure A2.8 shows all the objects that will be used to create the abstraction of this keyboard.

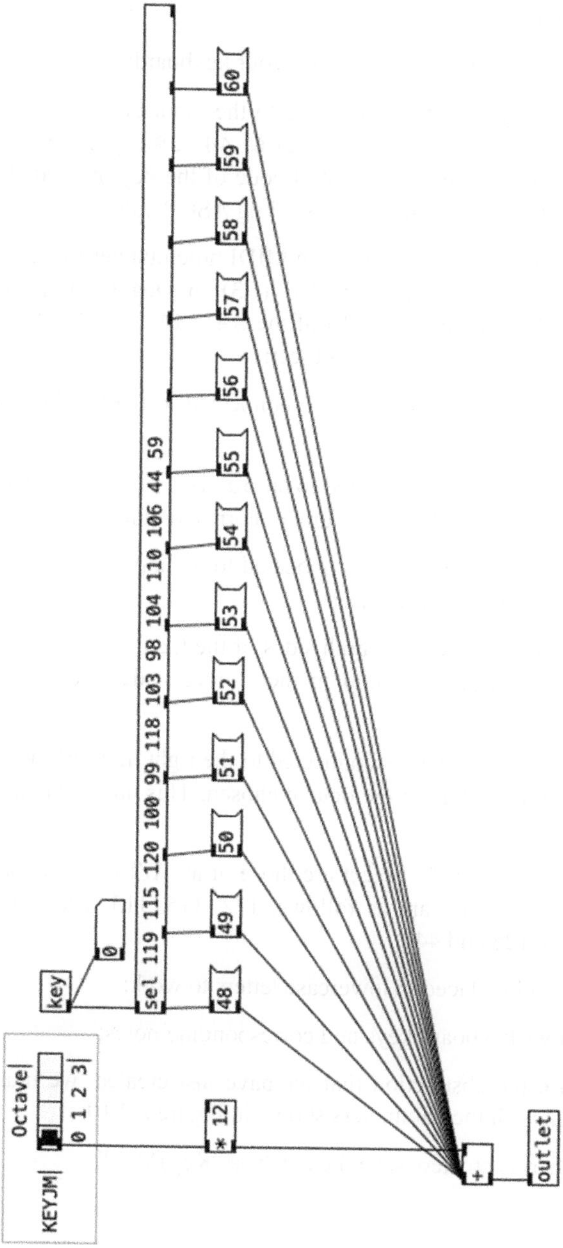

Figure A2.8. *The KeyJM abstraction patch, a simple virtual keyboard*

Instructions and comments are as follows:

– create a new patch;

– create a `key` object to capture the keys on your keyboard;

– the output of the `key` object is connected to the input of the `sel 119 115 120 100 99 118 103 98 104 110 106 44 59` object and the input of the `number` object, which controls the ASCII code of the key pressed. This object selects one of its outputs based on the corresponding ASCII value received;

– a set of 13 `message` objects defines the MIDI note number to be transmitted, from 48 to 60, that is, an octave from Do3 (C3) to Do4 (C4). Each input is connected independently to outputs 1 to 13 of the `sel 119 115 120 100 99 118 103 98 104 110 106 44 59` object;

– all outputs of the `message` objects are connected to the left input of the + object;

– the output of the + object is connected to an `outlet` object, which defines the data output of the abstraction (MIDI note number corresponding to a keyboard key);

– a `hradio` object with four cells is connected to the input of the `* 12` object, which defines an octave chosen by the user;

– a name and legends are added on both sides of the `hradio`, `KEYJM`, `Octave et 0 1 2 3` object. They are found in the `canvas`, defined below in these instructions;

– the output of the `* 12` object is connected to the right input of the + to define the MIDI note number according to the octave chosen. This number is increased by a multiple of 12 (0, 12, 24, or 36);

– the ASCII codes of the `sel` object are those of an AZERTY keyboard. For a QWERTY keyboard, the codes are as follows: 122, 115, 115, 120, 100, 99, 118, 103, 98, 104, 110, 106, 125 and 44;

– the keyboard must be placed in lowercase letters to work;

– Figure A2.9 shows keyboard keys and corresponding notes;

– to be able to use the abstraction that we have just created, we must define a `canvas` for the patch with the parameters shown in Figure A2.10;

– when the `canvas` is created, save the patch as “KeyJM.pd”;

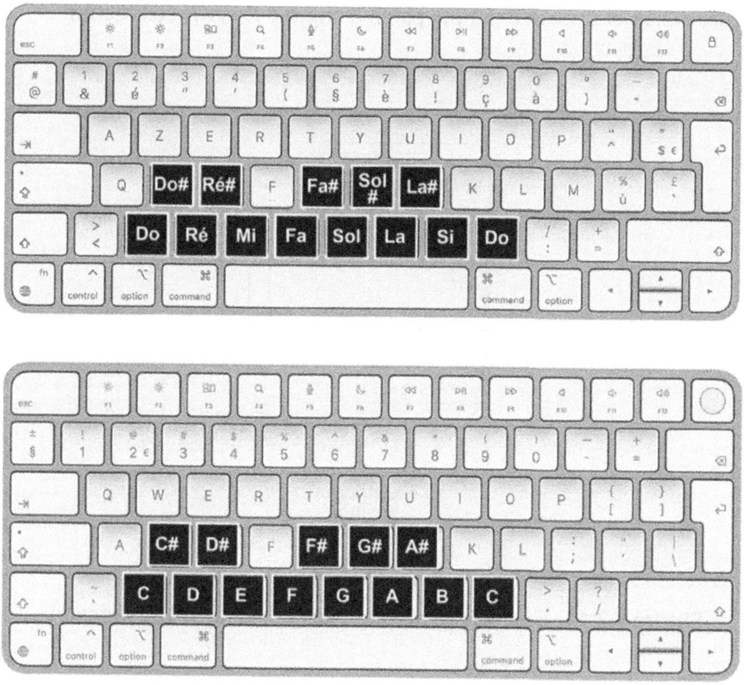

Figure A2.9. *Notes on an AZERTY and QWERTY keyboard*

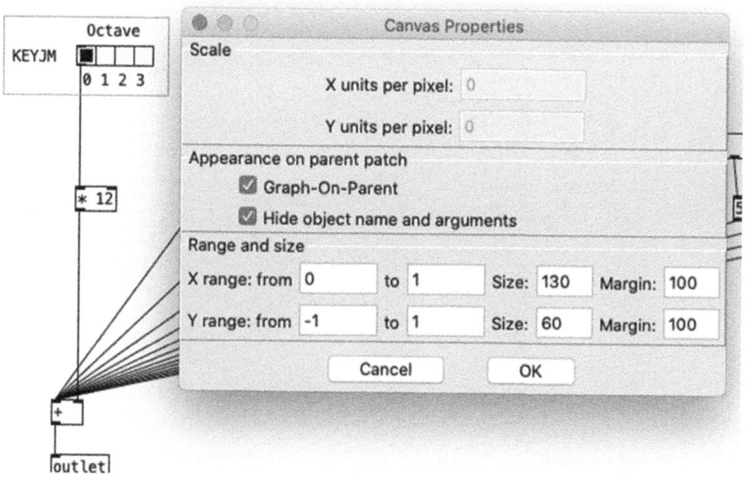

Figure A2.10. *Canvas properties of the patch*

– Figure A2.11 shows an example patch using the KeyJM abstraction;

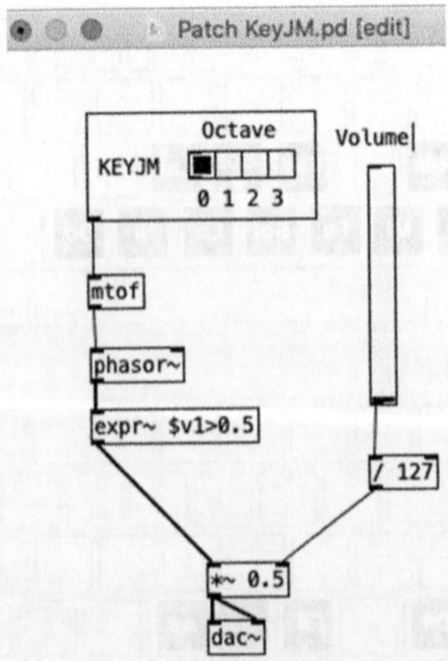

Figure A2.11. *Example of a patch using the KeyJM abstraction*

– the output of the abstraction is connected to the input to an mtof object. The rest of the patch is standard and generates a rectangular signal with adjustable volume.

Appendix 3

Keyboards and Interfaces

Using software synthesizers like the VCV Rack or Native Instruments Reaktor and visual programming languages like Max/MSP or Pure Data generally require a MIDI keyboard.

This keyboard can also be in the form of software. Max/MSP integrates one into its programming environment; as for Pure Data, Appendix 2 (see section A2.3) presents some solutions to meet this requirement.

There are several MIDI keyboards; some can be connected directly via USB, others use a MIDI port, and for some, both.

A MIDI connection to a microcomputer can only be made through an audio-MIDI or MIDI interface.

A3.1. MIDI keyboards

Figure A3.1 shows three examples of MIDI keyboards with their MIDI, MIDI and USB, and USB input–output ports.

MIDI master keyboards can integrate many features: a sequencer, an arpeggiator, pads, assignable buttons, etc.

We can often add pedals that can switch certain functions (damper and sustain pedals, etc.).

For a color version of all the figures in this chapter, see www.iste.co.uk/reveillac/synthesizers2.zip.

They do not all have bidirectional ports; in fact, a keyboard is more a device from which data is sent to a hardware or software synthesizer. However, some advanced keyboards include MIDI IN, MIDI OUT or USB ports that interact at both the input and output.

Figure A3.1. *MIDI keyboards and connections (Aturia Keystep, M-Audio Oxygen Pro 49, Studiologic SL73 Studio)*

Some keyboards have synchronization ports and can work with the CV/Gate standard (see Volume 1, Chapter 5).

A3.2. Audio-MIDI interface

No interface is necessary for keyboards that transmit MIDI data through a USB port. A simple driver, provided by the manufacturer and installed on your computer will do the trick, although this is not always essential.

For keyboards that only have a MIDI Out port, an audio or MIDI interface is warranted. On certain microcomputers, a sound card incorporating MIDI connectivity can act as an interface.

There are many audio interfaces offering a wide range of connections. They do not always require a MIDI port. This must be checked since this port can be a USB MIDI port or a classic MIDI port (five-pin DIN socket).

Figure A3.2. *MIDI and USB MIDI audio interface (MOTU Audio Express USB-Firewire)*

Different audio interfaces communicate with the microcomputer via different standards: USB, USB2, USB3, Firewire 400, Firewire 800 or Thunderbolt.

There are also interfaces dedicated only to the MIDI standard, which usually have two MIDI IN and MIDI OUT ports. These interfaces are inexpensive and reliably transmit all MIDI data, which is sufficient in most cases.

Figure A3.3. *Simple MIDI interface (M-Audio USB Uno)*

The connection to the microcomputer is made via USB. Interfaces are usually powered directly from the USB port.

Appendix 4

MIDI Notes, Numbers and Frequencies

In Table A4.1, you will find all the MIDI note numbers, their equivalent in piano key numbers, their American notation and their frequency.

The names of the notes differ depending on the country. For Latin and Slavic countries, it is DO or UT, RÉ, MI, FA, SOL, LA and SI, and for English-speaking countries, it is C, D, E, F, G, A and B.

Please note, for German-speaking countries, the letter B corresponding to the SI is replaced by the letter H (this difference is not considered in Table A4.1).

The reference frequency is A (A) at 440 Hz, and the middle note of a piano keyboard is C4 (C4).

MIDI note number	**Piano key number**	**English name**	**Latin name**	**Frequency (Hz)**
127	–	G9	SOL	12,543.85
126	–	F#9/Gb9	FA#9/SOLb9	11,839.82
125	–	F9	FA9	11,175.30
124	–	E9	MI9	10,548.08
123	–	D#9/Eb9	RÉ#9/MIb9	9,956.06
122	–	D9	RÉ9	9,397.27
121	–	C#9/Db9	DO#9/RÉb9	8,869.84
120	–	C9	DO9	8,372.02
119	–	B8	SI8	7,902.13
118	–	A#8/Bb8	LA#8/Bb8	7,458.62
117	–	A8	LA8	7,040.00

116	–	G#8/Ab8	SOL#8/LAb8	6,644.88
115	–	G8	SOL8	6,271.93
114	–	F#8/Gb8	FA#8/SOLb8	5,919.91
113	–	F8	FA8	5,587.65
112	–	E8	MI8	5,274.04
111	–	D#8/Eb8	RÉ#8/MIb8	4,978.03
110	–	D8	RÉ8	4,698.64
109	–	C#8/Db8	DO#8/RÉb8	4,434.92
108	88	C8	DO8	4,186.01
107	87	B7	SI7	3,951.07
106	86	A#7/Bb7	LA#7/Bb7	3,729.31
105	85	A7	LA7	3,520.00
104	84	G#7/Ab7	SOL#7/LAb7	3,322.44
103	83	G7	SOL7	3,135.96
102	82	F#7/Gb7	FA#7/SOLb7	2,959.96
101	81	F7	FA7	2,793.83
100	80	E7	MI7	2,637.02
99	79	D#7/Eb7	RÉ#7/MIb7	2,489.02
98	78	D7	RÉ7	2,349.32
97	77	C#7/Db7	DO#7/RÉb7	2,217.46
96	76	C7	DO7	2,093.00
95	75	B6	SI6	1,975.53
94	74	A#6/Bb6	LA#6/Bb6	1,864.66
93	73	A6	LA6	1,760.00
92	72	G#6/Ab6	SOL#6/LAb6	1,661.22
91	71	G6	SOL6	1,567.98
90	70	F#6/Gb6	FA#6/SOLb6	1,479.98
89	69	F6	FA6	1,396.91
88	68	E6	MI6	1,318.51
87	67	D#6/Eb6	RÉ#6/MIb6	1,244.51
86	66	D6	RÉ6	1,174.66
85	65	C#6/Db6	DO#6/RÉb6	1,108.73
84	64	C6	DO6	1,046.50
83	63	B5	SI5	987.77

82	62	A#5/Bb5	LA#5/Bb5	932.33
81	61	A5	LA5	880.00
80	60	G#5/Ab5	SOL#5/LAb5	830.61
79	59	G5	SOL5	783.99
78	58	F#5/Gb5	FA#5/SOLb5	739.99
77	57	F5	FA5	698.46
76	56	E5	MI5	659.26
75	55	D#5/Eb5	RÉ#5/MIb5	622.25
74	54	D5	RÉ5	587.33
73	53	C#5/Db5	DO#5/RÉb5	554.37
72	52	C5	DO5	523.25
71	51	B4	SI4	493,88
70	50	A#4/Bb4	LA#4/Bb4	466.16
69	49	A4	LA4	440.00
68	48	G#4/Ab4	SOL#4/LAb4	415.30
67	47	G4	SOL4	392.00
66	46	F#4/Gb4	FA#4/SOLb4	369.99
65	45	F4	FA4	349.23
64	44	E4	MI4	329.63
63	43	D#4/Eb4	RÉ#4/MIb4	311.13
62	42	D4	RÉ4	293.66
61	41	C#4/Db4	DO#4/RÉb4	277.18
60	40	C4	DO4	261.63
59	39	B3	SI3	246.94
58	38	A#3/Bb3	LA#3/Bb3	233.08
57	37	A3	LA3	220.00
56	36	G#3/Ab3	SOL#3/LAb3	207.65
55	35	G3	SOL3	196.00
54	34	F#3/Gb3	FA#3/SOLb3	185.00
53	33	F3	FA3	174.61
52	32	E3	MI3	164.81
51	31	D#3/Eb3	RÉ#3/MIb3	155.56
50	30	D3	RÉ3	146.83
49	29	C#3/Db3	DO#3/RÉb3	138.59

48	28	C3	DO3	130.81
47	27	B2	SI2	123.47
46	26	A#2/Bb2	LA#2/Bb2	116.54
45	25	A2	LA2	110.00
44	24	G#2/Ab2	SOL#2/LAb2	103.83
43	23	G2	SOL2	98.00
42	22	F#2/Gb2	FA#2/SOLb2	92.50
41	21	F2	FA2	87.31
40	20	E2	MI2	82.41
39	19	D#2/Eb2	RÉ#2/MIb2	77.78
38	18	D2	RÉ2	73.42
37	17	C#2/Db2	DO#2/RÉb2	69.30
36	16	C2	DO2	65.41
35	15	B1	SI1	61.74
34	14	A#1/Bb1	LA#1/Bb1	58.27
33	13	A1	LA1	55.00
32	12	G#1/Ab1	SOL#1/LAb1	51.91
31	11	G1	SOL1	49.00
30	10	F#1/Gb1	FA#1/SOLb1	46.25
29	9	F1	FA1	43.65
28	8	E1	MI1	41.20
27	7	D#1/Eb1	RÉ#1/MIb1	38.89
26	6	D1	RÉ1	36.71
25	5	C#1/Db1	DO#1/RÉb1	34.65
24	4	C1	DO1	32.70
23	3	B0	SI0	30.87
22	2	A#0/Bb0	LA#0/Bb0	29.14
21	1	A0	LA0	27.50

Table A4.1. *Corresponding MIDI number, note and frequency*

Glossary

Acoustic impedance: resistance of the medium (gas, liquid and solid) to the passage of an acoustic wave. Its unit is Pascal Second Per Meter (Pa s/m), also called a rayl.

ADSR (attack, decay, sustain, release): a set of parameters that form the envelope of a sound, that is, how it develops over time.

Aliasing: a phenomenon that introduces undesirable frequencies into a sampled signal or a signal modulating a carrier frequency. They appear when the carrier or sampling frequency is less than twice the maximum frequency of the signal.

Aliasing frequency: see aliasing.

Aperiodic signal: a signal whose form does not repeat itself, that is, that does not have a period.

Audio interface: sometimes called a sound card, it is a peripheral computer device that allows microphones, instruments and monitoring systems to be connected to a computer via a USB, PCI, Firewire or Thunderbolt port. It provides analog-to-digital conversion.

Auto-oscillation: also called self-oscillation. It is the generation and maintenance of periodic motion by a source devoid of corresponding periodicity. The oscillator itself controls the phase with which the external power supply acts on it.

Bandwidth: the frequency range of a sound source. We can also speak of sound spectrum width. It corresponds to the spectral size, that is, the interval between the low and high frequencies used by the source.

Beat: periodic modulation of an audio signal, consisting of two superimposed audio signals of close frequencies.

Bel: a logarithmic measurement unit of the ratio between two powers of an audio signal.

Bender: used to change the sound toward the bass or the treble. It can be done manually (wheel, joystick, etc.) or automatically via an envelope generator.

Binaural: is based on a sound capture method adapted to the morphology of the human head. This recording technique guarantees a very natural three-dimensional sound reproduction with excellent spatialization.

Brownian motion: a disordered and random movement, described mathematically, of a large particle immersed in a fluid or a gas, which is subjected to no other interaction than with the small molecules of the surrounding fluid.

Bypass: a mode that allows an audio signal to pass through a processing device without affecting its sound characteristics.

Carrier frequency: also called a carrier wave, or simply a carrier. It is a waveform that is modulated by an input signal (modulating signal or modulating frequency) in order to modify it (generally to apply an effect to it).

Carrier signal: see carrier frequency.

Celerity: also called speed. The speed of sound is the speed of propagation of sound waves in all media, gaseous, liquid or solid.

Click: high-amplitude, short-duration sound at the beginning of a waveform. It is found in music, noise or speech. It contains many non-periodic and high amplitude components at high frequencies. They often occur at the creation of a tone.

Combfilter: widely used in signal processing to add a delayed version of the signal to itself, which causes constructive or destructive interference.

Cutoff frequency: the limit frequency of useful operation of an electronic circuit, particularly for filters. High and low cutoff frequencies define the bandwidth.

CV/Gate (control voltage/gate): an electronic process for transmitting control information, used by many synthesizers and some other equipment such as sequencers or drum machines.

DAW: digital audio workstation.

DCO (digitally controlled oscillator): an oscillator whose frequency is digitally controlled. It is not a voltage that controls its frequency and amplitude but numerical values. The oscillator itself remains analog.

Decibel (dB): unit of magnitude, used in acoustics and electronics, defined as 10 times the decimal logarithm of the ratio between two powers.

Diaphony: interference between two audio signals.

Distortion: unwanted changes to an audio signal.

DO (digital oscillator): its frequency and waveform are created via software. It imitates an analog oscillator while eliminating or keeping its defects (stability over time).

Doppler effect: the frequency shift of an acoustic wave, observed between the measurements on transmission and reception when the distance between the transmitter and the receiver varies over time.

DSP (digital signal processing): a dedicated and optimized microprocessor to execute signal processing applications (filtering, extraction, conversion, encoding, decoding, compression, etc.).

Envelope: see ADSR.

Eurorack: a modular synthesizer format originally specified in 1996 by Doepfer Musikelektronik. It has made it possible to standardize the size and connectors of musical modules.

Firewire: trade name given by Apple to a multiplexed serial interface, also known by Sony under the name of i.Link, linked to the standard IEEE 1394.

Glide: ensures the transition from one note to another gradually. Most synthesizers have it; the passing speed between notes can be adjusted.

Grain: sound samples with a duration generally between 10 and 100 ms. These are the building blocks of granular sound synthesis. We control their density, pitch, length and envelope.

Haas effect: also called the "precedence effect," increases the stereophonic field. It is a psychoacoustic effect that uses a time shift imperceptible by the ear resulting in a greater field width.

Harmonic: also called a harmonic partial. It is a component of a periodic sound, the frequency of which is a multiple of the fundamental frequency. It can be even or odd.

Hertz: Hz is the International System (SI) unit of frequency. It is equivalent to one oscillation per second (s^{-1} or $^1/_s$).

HRTF (head-related transfer function): a transfer function characterized by the transformations brought to a sound signal by a listener's body, mainly the head and the ears. This allows humans to locate a sound in space, in azimuth (horizontally) and elevation (vertically).

ILD (interaural level difference): the intensity and frequency distribution difference between the two ears. Our ears can detect differences in loudness between the left and right ears, related to the acoustic shadow created by our head, which allows us to define the position of a sound source with a frequency higher than 1,500 Hz.

Interference: the combination of two sound waves of the same frequency. The interferences are the cause of diffraction and beat phenomena. The two types of interference, constructive and destructive, depend on the phase difference between the two sound waves.

ITD (interaural time difference): represents the difference in the arrival time of a sound wave at each ear. It is an important index to estimate the position of a sound source in the horizontal plane. It is effective in locating a source emitting waves whose frequency is lower than 1,500 Hz.

Key tracking: allows us to adapt the filter's cutoff frequency to the frequency of the note played. Thus, this cutoff frequency will no longer be fixed in relation to the entire sound spectrum but will vary for each note to obtain a more homogeneous rendering.

Keyboard splitter: divides the range of notes on a keyboard into several zones to assign them different sounds.

LFO (Low-frequency oscillator): a type of oscillator dedicated to low frequencies; its function is to control rather than to generate sound, that is, it can modulate another parameter.

Loudness: a numerical value representing the sound volume as perceived by a human being.

MIDI (Musical Instrument Digital Interface): protocol, communication standard and file format (MIDIfile) created for electronic musical instruments, which officially appeared in 1983. Its connectors are based on three five-pin DIN connectors: input (MIDI IN), output (MIDI OUT) and redirection (MIDI THRU). Two cables are required to ensure two-way communication between the various elements of the musical chain.

Modulation frequency: the frequency of one signal that modulates another. Modulation can be applied in frequency or amplitude.

Optocoupler: also called photo-coupler. It is an electronic component capable of transmitting the signal from one electrical circuit to another, without there being any galvanic contact between them.

Overdrive: name of an audio effect intended to recreate the saturation of an amplifier (usually an electric guitar) pushed to its gain limits.

Passband: the range of frequencies for which the attenuation of an audio signal is less than a specified value.

Phase: an instantaneous situation of a magnitude that varies cyclically. Two sound signals have different phases when they are of the same frequency and amplitude but are not superimposed.

Pitch-shifter: hardware or software tool, part of the sound effects, that allows for the modification of the pitch of a sound signal while remaining at the same tempo (without modifying its duration).

Presence: a device that adds treble and sharpness to sound. Generally, it acts directly on the amp's power stage via a feedback system (a system of reinjection of part of the signal at the amp's input).

Resonance: the tendency of an acoustic system to absorb more energy when the frequency of its oscillations reaches its natural frequency of vibration (resonant frequency).

Ring modulator: its principle is based on modulating the amplitude of a carrier wave by modulating a wave, which is often sinusoidal. If the modulating wave is audible (>20 Hz), new frequencies are obtained, replacing the carrier and modulating waves. These frequencies are the sum and the difference between the two previously mentioned waves.

Sample and hold: an electronic circuit consisting of a controlled switch and a capacitor. The purpose of a sample-and-hold is to sample an input signal at different times and then hold the last sample at a constant value.

Sampling: the digitization of a sound signal to create samples. It is carried out using a sampler.

Sound design: mastery and technical design of a soundtrack, within a framework defined upstream. The job of a sound designer is to develop a sound and artistic landscape that includes many elements, such as sound effects, music and artificial sounds.

Tessitura: set or scale of notes that the voice can emit without difficulty with the same sound volume and correct timbre.

Thunderbolt: connection and computer interface format designed by Intel, whose work began in 2007. The Mini Display Port connector, already present on Apple computers, was chosen as the standard interface for Thunderbolt. The latest versions (3 and 4) have a speed of 40 Gbit/s.

Timbre: sound parameters that characterize and identify an instrument or a voice.

Tonotopy: the spatial arrangement of the site where sounds of different frequencies are processed in the brain. Tonotopy in the auditory system begins in the cochlea, the small snail-shaped structure located in the inner ear, which sends information about the sound signal received by the ear to the brain.

Tremolo: periodic variation in the volume (and/or pitch) of a sound signal, usually produced by an LFO.

Universal serial bus (USB): the standard that defines a serial transmission linking several peripherals to a microcomputer or any other type of compatible device. It was designed in the mid-1990s and replaced many communication ports such as the parallel port, serial port and SCSI port. It is linked to a specific connection and has been able to evolve over time by increasing its speed, which can theoretically reach 10 Gbits/s in its latest version, USB 4, dating from 2017.

Vacuum tube: also called a "lamp" or "electronic tube". An active electronic component generally dedicated to the amplification of a signal. It was replaced later by semiconductors (transistor, diode, etc.). It was invented by Lee De Forest in 1907 (triode).

VCF (voltage-controlled filter): a module of an analog synthesizer including the filter part. This circuit is controlled by a voltage generated from a keyboard, sequencer or MIDI CV/Gate data.

VCO (voltage-controlled oscillator): a basic element of synthesis. It creates a waveform whose frequency and pitch are controlled by a voltage usually produced by a keyboard, sequencer or MIDI CV/Gate data.

Vibrato: periodic variation of the frequency of a sound signal generally produced by an LFO.

Waveshaping: an audio synthesis and transformation technique that transforms simple sounds into complex sounds. A pure sound, like a sine wave, can be transformed into a harmonically richer sound by changing its shape.

References

Anderton, C. (1986). *MIDI for Musicians*. The Music Sales Group, London.

Bianchi, F., Cipriani A., Giri, M. (2021). *Pure Data: Electronic Music and Sound Design – Theory and Practice – Volume 1*. ConTempoNet, Rome.

Braut, C. (1994). *Norme MIDI, Tome 1*. Sybex, Paris.

Braut, C. (1995). *Norme MIDI, Tome 2*. Sybex, Paris.

Chung, B. (2013). *Multiumedia Programming with Pure Data*. Packt Publishing, Birmingham.

Cohen-Levinas, D. (1999). *La synthèse sonore*. Éditions du Centre Pompidou, Paris.

Collins, N., Schedel, M., Wilson, S. (2013). *Electronic Music*. Cambridge University Press, Cambridge.

De Wilde, L. (2016). *Les fous du son*. Grasset, Paris.

Friedman, D. (1985). *Complete Guide to Synthesizers, Sequencers, and Drum Machines*. Music Sales Corp, California.

Grosse, D. (2021a). *Modular Synthesizer Mastery – Volume 1*. 20 Objects Publishing, Northfield.

Grosse, D. (2021b). *Modular Synthesizer Mastery – Volume 2*. 20 Objects Publishing, Northfield.

Grosse, D. (2021c). *Modular Synthesizer Mastery – Volume 3*. 20 Objects Publishing, Northfield.

Hillerson, T. (2014). *Programming Sound with Pure Data*. O'Reilly, Massachusetts.

Kendall, G. and Puddie Rodgers, C.A. (1981). The simulation of three-dimensional localization cues for headphone listening. *Proceedings of the International Computer Music Conference*, Denton, Texas.

Keyboard Magazine (1985). *Synthesizers and Computers*. Hal Leonard, Winona.

Lechner, P. (2014). *Multimedia Programming Using Max/MSP and TouchDesigner*. Packt Publishing, Birmingham.

Lehmann, H. (2017). *La Révolution digitale dans la musique*. Allia, Paris.

Manning, P. (2013). *Electronic and Computer Music*. OUP USA, Oxford.

Manzo, V.J. (2016). *Max/MSP/Jitter for Music: A Practical Guide to Developing Interactive Music Systems for Education and More*. OUP USA, Oxford.

Mercier, D. (ed.) (2002). *Le livre des techniques du son*. Dunod, Paris.

Meyer, C. and Brooks, E. (1994). MIDI time code, detailed specification. *Music Technology*.

Pierce, J.R. (1984). *Le son musical*. Belin, Paris.

Pinch, T. and Trocco, F. (2002). *Analog Days: The Invention and Impact of the Moog Synthesizer*. Harvard University Press, Cambridge, Massachusetts.

Quinet, J.J. (1987). *Les cahiers de l'ACME. Le système MIDI*. Jean Jacques Quinet Éditeur, France.

Rausch, K. (2021). *The DX Story: FM Synthesis – The Magic Formula for the Sound of 80s*. NeoPubli GmbH, Berlin.

Réveillac, J.-M. (2017). *Musical Sound Effects: Analog and Digital Sound Processing*. ISTE Ltd, London, and John Wiley & Sons, New York.

Roads, C. (2015). *Composing Electronic Music: A New Aesthetic*. OUP USA, Oxford.

Roads, C. (2016). *L'audionumérique – Musique et informatique*, 3rd edition. Dunod, Paris.

Rothstein, J. (1995). *MIDI: A Comprehensive Introduction – Volume 7*. A-R Editions, Wisconsin.

Taylor, G. (2018). *Step by Step: Adventures in Sequencing with Max/MSP*. Cycling'74, California.

Vail, M. (2000). *Vintage Synthesizers*. Backbeat Books, Connecticut.

Vail, M. (2014). *The Synthesizer: A Comprehensive Guide to Understanding, Programming, Playing, and Recording the Ultimate Electronic Music Instrument*. OUP USA, Oxford.

Internet links

Internet links are inherently volatile. They may evolve over time into other addresses, or even disappear. All were valid when writing this book; if some of them no longer work, a little research on Google or with your favorite search engine will help you find them.

Software manufacturers and publishers

Ableton: www.ableton.com

Akai: www.akaipro.com

Alesis: www.alesis.com

Arturia: www.arturia.com

Behringer: www.behringer.com

CTRLR: www.ctrlr.org

Cycling'74: www.cycling74.com

Dave Smith: www.davesmithinstruments.com

Doepfer: www.doepfer.de

E-mu: www.emu.com

Gforce Software: www.gforcesoftware.com

IK Multimedia: www.ikmultimedia.com

Korg: www.korg.com

Moog: www.moogmusic.com

MOTU: www.motu.com

Native Instruments: www.native-instruments.com

Novation: www.novationmusic.com

Synapse Audio Software: www.synapse-audio.com

U-He: www.u-he.com

Universal Audio: www.uaudio.com

VCV Rack: www.vcvrack.com

Yamaha: www.yamaha.com

Vintage electronic instruments

Synthmuseum, the website for old synthesizers: www.synthmuseum.com

Synthtopia, the museum for old synthesizers: www.synthtopia.com

The 14 most important synthesizers in electronic music: www.factmag.com/2016/09/15/14-most-important-synths

Vintage synthesizers: www.vintagesynth.com

Technical documentation

A website dedicated to the technical documentation of synthesizers: www.synthxl.com

A website dedicated, among other things, to the technical documentation of synthesizers: www.synthfool.com

General sites

A vintage synthesizers website: www.vintagesynth.com

Attack magazine, a music and digital audio magazine: www.attackmagazine.com

Audiofanzine, a website dedicated to music audio equipment: www.audiofanzine.com

Canford, a seller of audio and video products (in French): www.canford.fr

Harmony Central, one of the best websites dedicated to music audio: www.harmonycentral.com

Keyboard magazine, a music magazine: www.keyboardmag.com

KR Home studio, the music creation magazine (in French): www.kr-homestudio.fr

Music equipment: www.musicradar.com

Music equipment and more: www.pmtonline.co.uk

Music, musicians and instruments: www.factmag.com

Music Store Professional, seller of equipment: www.musicstore.de/en_OE/EUR

MusicTech, a website for sound engineers and for music production: www.musictech.net

Musiker Board, a music audio website (in German): www.musiker-board.de

Pro Audio Review magazine: www.prosoundnetwork.com/article.aspx?articleid=39995

ProSound: www.prosoundnetwork.com

SoundClick, a music audio website: www.soundclick.com

Tape Op magazine, a music magazine: www.tapeop.com

Thomann, a seller of online music products: https://www.thomann.de/gb/index.html

Training and resources for sound lovers: www.deveniringeson.com

Woodbrass, an online music seller: www.woodbrass.com

Zikinf, a general website for music and audio equipment: www.zikinf.com

Interfaces and communication

Digital audio connections (in French): www.mamosa.org/jenfi.home/debuter/connexions numeriques.php

Digital interfaces: www.soundonsound.com/techniques/digital-interfacing

Max/MSP

Ableton and Max for Live website: www.ableton.com/en/live/max-for-live

Cycling'74 publisher website: www.cycling74.com

Max objects database: www.maxobjects.com

Max/MSP course, introduction to Max/MSP: www.instructables.com/Intro-to-MaxMSP

Max/MSP course, vision and sound with Max: www.goldbergs.com/max/parsons

Max/MSP forum: www.cycling74.com/forums/page/1

Patcher for Max/MSP, Max/MSP Codelab: www.codelab.fr/max-msp

vmpk, virtual MIDI keyboard: www.vmpk.sourceforge.io/#Introduction

MIDI

An introduction to the MIDI standard: www.midi.org/articles-old/an-intro-to-midi

An introduction to the MIDI standard (in French): www.cri.ensmp.fr/~pj/music_slides.pdf

Differences between GM, GS and XG: www.cybermidi.com/helpdesk/knowledgebase.php?article=47

Exploring the GM standard: www.harfesoft.de/aixphysik/sound/midi/pages/genmidi.html

General MIDI (in French): www.daffyduke.lautre.net/zik/midi_10.html

MIDI 2.0: www.midi.org/midi-articles/details-about-midi-2-0-midi-ci-profiles-and-property-exchange

MIDI forum: www.midi.org/forum/830-midi-octave-and-note-numbering-standard

The GM standard: www.midi.org/specifications-old/item/gm-level-1-sound-set

The MIDI standard (in French): www.ntemusique.free.fr/musique/MIDI/MIDI.pdf

The MIDI standard (in French): www.ogloton.free.fr/midi/presentation.html

The MIDI standard (in French): www.sonelec-musique.com/electronique_theorie_midi_norme.html

The MIDI standard and its files (in French): www.jchr.be/linux/midi-format.htm

The MIDI XG format (in French): www.ppretot.free.fr/whatxgf.htm

Pure Data

Noisy box, Pure Data Abstractions: https://noisybox.net/pd

pad.scope, an oscilloscope for Pure Data: www.patchstorage.com/pad-library

Pure Data: www.puredata.info

Pure Data Forum: www.forum.pdpatchrepo.info

Pure Data GitHub: www.github.com/pure-data/pure-data

Resources for Pure Data: www.mortmain.com/pd.html

Resources for Pure Data, Floss Manuals Pure Data: www.fr.flossmanuals.net/puredata/introduction

Virtual keyboard

A piano keyboard in Pure Data from your AZERTY keyboard: www.puredata.info/Members/anfex/teclado/view?searchterm=keyboard

A virtual MIDI keyboard in Pure Data: www.puredata.info/Members/fr4nck/fr4-azerty.pd/view?searchterm=keyboard

Keybin, a MIDI keyboard in Pure Data from your microcomputer keyboard: www.puredata.info/Members/sokratesla/abstractions/keybin.pd/view?searchterm=keyboard

vmpk, the virtual MIDI keyboard for MacOS, Microsoft Windows and Linux: www.vmpk.sourceforge.io/#Introduction

Reaktor blocks

Modules for Reaktor – Toybox: www.toyboxaudio.com

Native Instruments Blocks Wired: www.audiopluginsforfree.com/native-instruments-blocks-wired/

Tutorial – Getting started with Reaktor blocks: www.ask.video/article/audio-software/getting-started-with-reaktor-blocks

Tutorial – How to build a synth in Reaktor: www.adsrsounds.com/reaktor-tutorials/how-to-build-a-synth-in-reaktor-a-beginners-guide

Tutorial – How to build your first Reaktor synth: www.musicradar.com/news/how-to-build-a-reaktor-synth

Video tutorial – Reaktor know-how – Blocks: www.groove3.com/tutorials/REAKTOR-Know-How-Blocks

Working with Reaktor Blocks – Processing audio with Reaktor Blocks: www.macprovideo.com/article/audio-software/processing-audio-with-ni-reaktor-blocks

Software synthesizers

Arturia ARP 2600 V: www.arturia.com/products/analog-classics/arp2600v

Best virtual Minimoogs: www.careersinmusic.com/moog-vst

Best VST/AU Minimoog emulation: www.musicradar.com/news/tech/6-of-the-best-vst-au-minimoog-emulation-plugins-622621

Cherry Audio CA2600: www.cherryaudio.com/products/ca2600

CTRLR, control your MIDI life –MIDI publisher: www.ctrlr.org

MiniMogueVA: www.plugins4free.com/plugin/405

MiniMoog, Arturia Mini V: www.arturia.com/products/analog-classics/mini-v/overview

Native Instruments Reaktor Blocks: www.native-instruments.com/fr/products/komplete/synths/reaktor-6/blocks

Sonivox TimeWarp 2600: www.sonivoxmi.com/virtual-instruments/time-warp-2600.html

VCV Rack: www.vcvrack.com

VCV Rack – Rack Github: www.github.com/VCVRack/Rack

VST emulation of famous synths: www.productionmusiclive.com/blogs/news/top-5-free-emulations-of-famous-synths

Sound synthesis

A granular synthesis resource: www.granularsynthesis.com/index.php

A slideshow on reverb: www.cs.wellesley.edu/~cs203/lecture_materials/reverb/reverb.pdf

ADSRSounds – Theory behind FM synthesis: www.adsrsounds.com/fm8-tutorials/theory-behind-fm-synthesis/

An introduction to synth programming for electronic music producers: www.renegadeproducer.com/audio-synthesis.html

AudioKeys – Sound synthesis and synthesizers forum (in French): www.audiokeys.net/forum

AudioShapers – Noise, types and colors: www.audioshapers.com/blog/noise-types-and-colors.html

Different types of noise: www.emastered.com/fr/blog/different-types-of-noise

FM synthesis (in French): www.inrp.fr/JIPSP/phymus/m_techni/synthfm/ac_syfm.htm

Freeverb: www.dsprelated.com/freebooks/pasp/Freeverb.html

Granular synthesis: www.sfu.ca/~truax/gran.html

History of sound synthesis (in French): www.frank-lovisolo.fr/WordPress/synthese-sonore-histoire-raccourci

Izotope – Basics of FM synthesis: www.izotope.com/en/learn/simple-fm-synthesis-sine-waves-and-processors.html

Izotope – The basics of granular synthesis: www.izotope.com/en/learn/the-basics-of-granular-synthesis.html

Music-tech – Learning the base of FM synthesis: www.musictech.com/guides/essential-guide/how-fm-synthesis-works/

Musical signal processing with LabVIEW: www.cnx.org/contents/4ODlu9s3@1.6:H3hx-Qm8@3/Schroeder-Reverberator

Natural sounding artificial reverberation: www.docplayer.net/58419605-Natural-sounding-artificial-reverberation.html

Noise and colors (in French): www.easyzic.com/dossiers/les-bruits,h391.html

Principles of sound synthesis (in French): www.easyzic.com/dossiers/principes-de-la-synthese, h392.html

Processing of audio signals: www.pact.wp.imt.fr/files/2011/09/EffetsetReverbGrichard.pdf

Schroeder reverberators: www.ccrma.stanford.edu/~jos/pasp/Schroeder_Reverberators.html

Schroeder reverberators: www.dsprelated.com/freebooks/pasp/Schroeder_Reverberators.html

Sound on Sound – Granular synthesis: www.soundonsound.com/techniques/granular-synthesis

Sound synthesis (in French): www.traitement-signal.com/synthese_sonore.php

Sound synthesis part 1: www.soundonsound.com/techniques/sound-synthesis-part-1

Sound synthesis part 2: www.soundonsound.com/techniques/sound-synthesis-part-2

SoundFly – An introduction to FM synthesis: www.flypaper.soundfly.com/produce/an-introduction-to-fm-synthesis/

Synthesizers (in French): www.frank-lovisolo.fr/WordPress/synthese-sonore-histoire-raccourci

The basics of sound synthesis: www.theproaudiofiles.com/sound-synthesis-basics

The beginning of sound synthesis: www.digitalsoundandmusic.com/6-1-1-the-beginnings-of-sound-synthesis/

Understanding synthesizers: www.splice.com/blog/understanding-synthesizers/

Various video tutorials

Elephorm: www.elephorm.com/audio-mao.html

Learning modular – Courses on modular synthesizers: www.learningmodular.com

Tutorom (in French): www.tutorom.fr/categories-de-tutoriels/fr/audio

Virtual production school (VPS) (in French): www.tutoriels-mao.com/les-tutoriels/mix-et-master-de-a-%C3%A0-z-avec-des-plugins-gratuits-detail

Index

P, R

S, T

U, V, W

Other titles from

iSTE

in

Waves

2022

GONTRAND Christian
Electromagnetism: Links to Special Relativity

LACAZE Pierre Camille, LACROIX Jean-Christophe
Nanotechnology and Nanomaterials for Energy

RÉVEILLAC Jean-Michel
Recording and Voice Processing 1: History and Generalities
Recording and Voice Processing 2: Working in the Studio

ROYER Daniel, VALIER-BRASIER Tony
Elastic Waves in Solids 1: Propagation

SAKHO Ibrahima
Nuclear Physics 1: Nuclear Deexcitations, Spontaneous Nuclear Reactions

VALIER-BRASIER Tony, ROYER Daniel
Elastic Waves in Solids 2: Radiation, Scattering, Generation

2021

DAHOO Pierre-Richard, LAKHLIFI Azzedine
Infrared Spectroscopy of Symmetric and Spherical Top Molecules for Space Observation 1
(Infrared Spectroscopy Set – Volume 3)
Infrared Spectroscopy of Symmetric and Spherical Top Molecules for Space Observation 2
(Infrared Spectroscopy Set – Volume 4)

2020

DANIELE Vito G., LOMBARDI Guido
Scattering and Diffraction by Wedges 1: The Wiener-Hopf Solution - Advanced Applications
(Waves and Scattering Set – Volume 1)
Scattering and Diffraction by Wedges 2: The Wiener-Hopf Solution - Advanced Applications
(Waves and Scattering Set – Volume 2)

SAKHO Ibrahima
Introduction to Quantum Mechanics 2: Wave-Corpuscle, Quantization & Schrödinger's Equation

2019

BERTRAND Pierre, DEL SARTO Daniele, GHIZZO Alain
The Vlasov Equation 1: History and General Properties

DAHOO Pierre-Richard, LAKHLIFI Azzedine
Infrared Spectroscopy of Triatomics for Space Observation
(Infrared Spectroscopy Set – Volume 2)

RÉVEILLAC Jean-Michel
Electronic Music Machines: The New Musical Instruments

ROMERO-GARCIA Vicente, HLADKY-HENNION Anne-Christine
Fundamentals and Applications of Acoustic Metamaterials: From Seismic to Radio Frequency
(Metamaterials Applied to Waves Set – Volume 1)

SAKHO Ibrahima
Introduction to Quantum Mechanics 1: Thermal Radiation and Experimental Facts Regarding the Quantization of Matter

2018

SAKHO Ibrahima
Screening Constant by Unit Nuclear Charge Method: Description and Application to the Photoionization of Atomic Systems

2017

DAHOO Pierre-Richard, LAKHLIFI Azzedine
Infrared Spectroscopy of Diatomics for Space Observation
(Infrared Spectroscopy Set – Volume 1)

PARET Dominique, HUON Jean-Paul
Secure Connected Objects

PARET Dominque, SIBONY Serge
Musical Techniques: Frequencies and Harmony

RÉVEILLAC Jean-Michel
Analog and Digital Sound Processing

STAEBLER Patrick
Human Exposure to Electromagnetic Fields

2016

ANSELMET Fabien, MATTEI Pierre-Olivier
Acoustics, Aeroacoustics and Vibrations

BAUDRAND Henri, TITAOUINE Mohammed, RAVEU Nathalie
The Wave Concept in Electromagnetism and Circuits: Theory and Applications

PARET Dominique
Antennas Designs for NFC Devices

PARET Dominique
Design Constraints for NFC Devices

WIART Joe
Radio-Frequency Human Exposure Assessment

2015

PICART Pascal
New Techniques in Digital Holography

2014

APPRIOU Alain
Uncertainty Theories and Multisensor Data Fusion

JARRY Pierre, BENEAT Jacques N.
RF and Microwave Electromagnetism

LAHEURTE Jean-Marc
UHF RFID Technologies for Identification and Traceability

SAVAUX Vincent, LOUËT Yves
MMSE-based Algorithm for Joint Signal Detection, Channel and Noise Variance Estimation for OFDM Systems

THOMAS Jean-Hugh, YAAKOUBI Nourdin
New Sensors and Processing Chain

TING Michael
Molecular Imaging in Nano MRI

VALIÈRE Jean-Christophe
Acoustic Particle Velocity Measurements using Laser: Principles, Signal Processing and Applications

VANBÉSIEN Olivier, CENTENO Emmanuel
Dispersion Engineering for Integrated Nanophotonics

2013

BENMAMMAR Badr, AMRAOUI Asma
Radio Resource Allocation and Dynamic Spectrum Access

BOURLIER Christophe, PINEL Nicolas, KUBICKÉ Gildas
Method of Moments for 2D Scattering Problems: Basic Concepts and Applications

GOURE Jean-Pierre
Optics in Instruments: Applications in Biology and Medicine

LAZAROV Andon, KOSTADINOV Todor Pavlov
Bistatic SAR/GISAR/FISAR Theory Algorithms and Program Implementation

LHEURETTE Eric
Metamaterials and Wave Control

PINEL Nicolas, BOURLIER Christophe
Electromagnetic Wave Scattering from Random Rough Surfaces: Asymptotic Models

SHINOHARA Naoki
Wireless Power Transfer via Radiowaves

TERRE Michel, PISCHELLA Mylène, VIVIER Emmanuelle
Wireless Telecommunication Systems

2012

LALAUZE René
Chemical Sensors and Biosensors

LE MENN Marc
Instrumentation and Metrology in Oceanography

LI Jun-chang, PICART Pascal
Digital Holography

2011

BECHERRAWY Tamer
Mechanical and Electromagnetic Vibrations and Waves

BESNIER Philippe, DÉMOULIN Bernard
Electromagnetic Reverberation Chambers

GOURE Jean-Pierre
Optics in Instruments

GROUS Ammar
Applied Metrology for Manufacturing Engineering

LE CHEVALIER François, LESSELIER Dominique, STARAJ Robert
Non-standard Antennas

2010

BEGAUD Xavier
Ultra Wide Band Antennas

MARAGE Jean-Paul, MORI Yvon
Sonar and Underwater Acoustics

2009

BOUDRIOUA Azzedine
Photonic Waveguides

BRUNEAU Michel, POTEL Catherine
Materials and Acoustics Handbook

DE FORNEL Frédérique, FAVENNEC Pierre-Noël
Measurements using Optic and RF Waves

FRENCH COLLEGE OF METROLOGY
Transverse Disciplines in Metrology

2008

FILIPPI Paul J.T.
Vibrations and Acoustic Radiation of Thin Structures

LALAUZE René
Physical Chemistry of Solid-Gas Interfaces

2007

KUNDU Tribikram
Advanced Ultrasonic Methods for Material and Structure Inspection

PLACKO Dominique
Fundamentals of Instrumentation and Measurement

RIPKA Pavel, TIPEK Alois
Modern Sensors Handbook

2006

BALAGEAS Daniel *et al.*
Structural Health Monitoring

BOUCHET Olivier *et al.*
Free-Space Optics

BRUNEAU Michel, SCELO Thomas
Fundamentals of Acoustics

FRENCH COLLEGE OF METROLOGY
Metrology in Industry

GUILLAUME Philippe
Music and Acoustics

GUYADER Jean-Louis
Vibration in Continuous Media